Watershed Development Inputs and Social Change

Watershed Development Inputs and Social Change

Understanding the Changing Culture of Child Nutrition

APOORVAA PANDIT

WATERSHED ORGANISATION TRUST (WOTR)
Ahmednagar, Maharashtra, India

RAWAT PUBLICATIONS
Jaipur • New Delhi • Bangalore • Hyderabad • Guwahati

ISBN 81-316-0342-3

This study was supported by GTZ (German Agency for Technical Cooperation)

Published by
Prem Rawat for **Rawat Publications**
Satyam Apts, Sector 3, Jawahar Nagar, Jaipur 302 004 (India)
Phone: 0141 265 1748 / 7006 Fax: 0141 265 1748
E-mail: info@rawatbooks.com
Website: www.rawatbooks.com

New Delhi Office
4858/24, Ansari Road, Daryaganj, New Delhi 110 002
Phone: 011 2326 3290

Also at Bangalore, Hyderabad and Guwahati

Typeset by Rawat Computers, Jaipur
Printed at Chaman Enterprises, New Delhi

Contents

Foreword

With 2015 just five years away, to achieve the Millennium Development Goal 1 Target 2, "*Halve the number of people suffering from hunger*" Indicator 4, "*Prevalence of underweight children under five years of age*" still poses a challenge. On the one hand, efforts are being made by government to achieve this through the ICDS that since long has been focusing on the perinatal period and children of the 0–6 years of age, through the *Anganwadi* outreach programme. On the other, government and various developmental agencies are investing heavily enhancing land productivity through soil and water conservation along watershed lines and agriculture improvement, thus increasing water and food availability. Despite these enabling conditions, we still lag behind with 47 per cent of our rural children falling in the 'below normal' weight-for-age category.

One possible reason for this unhappy state of affairs is that each activity or scheme at the village level is implemented independently of the other, despite the fact that the target population is the same. Desired outcomes can only be achieved if these schemes work synergistically with one another. That is to say, convergence of objectives and activities is the need of the hour. However, this at best can only be an externality efficiently leveraged. It is of far greater importance to address the cognitive and mindset of the villages together with underlying social and cultural structures, if desired and sustainable outcomes are to be realised. This was the purpose of initiating the "Participatory Child Growth Monitoring Project" by WOTR (Watershed Organisation Trust), Ahmednagar.

The approach developed by WOTR was not only to leverage existing and on-going schemes, but also and primarily, to bring about a mindset and cultural change in the way the villagers perceived the nutritional status of their children.

This publication is the first of a series of studies that seeks to answer the following questions: Do increasing income levels, resulting from successfully implemented watershed projects, translate into improved nutritional status for children in the 0–5 age group bracket? Should watershed development projects specifically include nutrition and growth monitoring inputs to raise the nutritional status of children? Should the current approach of the ICDS be supplemented with a strategy aimed at changing mindsets, practices and cultural patterns?

The present book results out of a detailed micro study of three villages in drought-prone Ahmednagar district in Maharashtra. It is one of a series of comparative and impact assessment studies of three villages each having one of the following characteristics: (i) with input of watershed development (ii) with inputs of watershed development plus growth monitoring of the 0–5 years children and (iii) with no inputs. A comparison of the impacts of the development inputs was the objective. All three villages have a mixed population of caste and tribal communities. The development impacts on both these communities are discussed at length. In particular, the process of cognitive, perceptual and cultural change in respect to child health and nutrition as an impact result of the approach adopted has been traced out in detail. For purposes of this study, a specific "CONCEPT POINTS SCALE" to measure and compare mindset and cultural changes across groups and communities has been developed.

This study convincingly establishes that unless specific inputs in regard to nutrition are included as part of watershed development, the increase in agriculture output, does not necessarily translate into improved nutrition of the local inhabitants as reflected in the most vulnerable group, the 0–5 years old children. It is therefore recommended that one of the key indicators of success of a watershed project be the extent of improvement of the nutritional status of children. Furthermore, desired outcomes and sustained benefits can only be achieved through dovetailing, convergence and synergising of related developmental interventions at the village level, in particular the ICDS and the watershed/agriculture development programmes.

Marcella D'Souza
Executive Director
WOTR

Preface

Health is an aspect of culture. Health perceptions and practices are governed by culture. Introduction of new technology, ideas, activities are likely to lead to a change in one or more parts of culture. A change in one part causes a change in other parts of culture. In the analysis of social change and development, changes surrounding health are rarely discussed. As health is a means and an end to development, documentation of health related changes in culture helps understand the impacts of development and provide a direction to undertake and continue with the kind of development work that is urgent and essential for sustainability. 'Child Nutrition' is the subject matter here for the analysis of social and health change as triggered by "Watershed Development Inputs". Development inputs are one of the many factors that play a significant role in changing people, changing culture.

The work is based on a micro study of social change, understanding the impact of two kinds of watershed development inputs: one is integrated watershed development and the other is NGO-initiated participatory growth monitoring input for children 0–5 years. It is clarified here that growth monitoring is a development input and includes different components besides recording of body measurements.

A large-scale study employing quantitative and qualitative social science research methods was designed. The quantitative study examined the nutritional status of children 0–5 years while the qualitative study focused on understanding the nutrition and health consciousness of the community.

The current script is based on qualitative study. The study explains social and health culture change in a comparative perspective. The comparison is made at three levels, at the level of two input groups and a no input group.

Large amount of qualitative data were collected from the field. Qualitative data analysis is a Herculean task for a social science researcher and more so when it is a comparative study at different levels: at three levels of inputs on one side and at the level of communities on the other. Data analyses were carried out manually into many steps with refinement. Analytical data are presented in the chapters. An attempt has been made towards quantification of the qualitative data for simplified understanding of abstract keeping development professionals in view. Quantitative parameters, viz., matrices, scores, scale have been used to treat the qualitative data as a part of analysis.

The cohabitation of Hindu caste and the tribal communities in the villages made community comparison an essential task. The path of development of the caste and the tribal communities, cross-cultural variations affecting livelihood of the poor, change among women and gender issues, the difference in the ideal and the actual behaviour are discussed in the light of child nutrition and health.

Identification of significant inputs contributing to change and grey areas of watershed development are put forth as considerations for accelerating the path of sustainable development towards nutrition/health ends.

The lessons learnt from the study draw attention to policy and research in the area of watershed development.

Apoorvaa Pandit

Acknowledgements

The study and the writings could be completed due to the support, cooperation and involvement of many individuals at various stages. I would like to thank all of them for their help.

I would like to gratefully acknowledge Dr Marcella D' Souza, Executive Director, Watershed Organisation Trust, Ahmednagar, who provided the direction for the study. I had her support and guidance throughout. She read all of the manuscript and made helpful suggestions. It is our privilege to have Dr Shaunak Kulkarni, Reader, Department of Anthropology, University of Pune and Dr Chandrakala Mannurru, Reader and Head, Department of Food Science and Nutrition, S.N.D.T. College, Pune as Research Advisors who provided valuable guidance whenever required.

Mr Dipak Zade, my Research Associate has been of great help to me on many fronts. In this particular study, he helped me in review of literature, data collection and organisation, and coordination with people in the villages. His efforts are greatly acknowledged.

I owe special thanks to senior officials from WOTR and SAMPADA TRUST, Ahmednagar: Mr Ganesh Rajapure, for inputs in selection of study villages and field coordination, Mr Joseph Shinde, for nominating students of 'Rural Chroniclers' course of WOTR for data collection, Ms Chayya Nikrad for providing data from WOTR database as and when required, Ms Preetilata Gaikwad for field inputs, and assistance by Ms Ujjwala Kavthekar and Ms Bhagyashree Shinde is highly appreciated.

I am thankful to Mr Kishor Telang, Mr Prasad Dharmadhikari from the Head Office, Mr Date and staff from Darewadi Training Centre, Darewadi, Ahmednagar and Mr Ajay Shelke from WOTR Research Unit, Pune for administrative support throughout the study.

Many thanks to my colleagues at the Research Unit, Pune: Mr Atul Pagar for photographs, Mr Chandrakant Deokar for suggestions from the point of view of agriculture and environment field and Mr Santosh Ramane for computer assistance. Thanks to Mr Sanjay Kulkarni for translation of study material, Ms Aparna Chaudhari for research assistance, Ms Sindhu Dhondge, Ms Savita Baris, Mr Navnath Ighe, Mr Swapnil Bhilkar, Mr Kailash Lokhande for data collection on village profiles.

I am grateful to the *Anganwadi* workers of the villages for their immense cooperation and support in our fieldwork as always. My heartfelt thanks to all the people from the villages. Their hospitality, accommodation arrangements, spending time answering our questions and all other support was overwhelming. Special thanks are due in this regard to Ms Nandatai Kale, Ms Sangeeta Bodke, health promoters and Mr Dilip Satkar from village Mhaswandi, Ahmednagar.

My sincere thanks are due to the drivers, WOTR, Mr D.G. Kakade, Mr Johnny Prabhune, Mr Sunil Walke, Mr Carlos Ohal for their help in commutation to the field at any time even at odd hours; Mr Abaso Katkar, Research Unit, for office assistance.

List of Tables

List of Figures and Plates

Figures

Plates

Introduction

Watershed Development Inputs

Watershed development is widely recognised as an effective means for poverty alleviation in rainfed rural areas. A widely stated cause of poverty in rain dependent rural areas is the low productivity of human and livestock resources coupled with deterioration of land and environmental resources. Comprehensive and integrated watershed development programmes are implemented with a focus on productivity enhancement and livelihood promotion alongside conservation and regeneration of natural resources.

Initiation of participatory growth monitoring of children 0–5 years in watershed villages steers child health development.

Maharashtra, a state in western India, falls in semi-arid or arid zone. A large part of it is drought-prone. Having experienced the recurrent droughts, an initiative towards poverty reduction through watershed development was taken up by the formation of Indo-German Watershed Development Programme (IGWDP) conceived and initiated by Fr. Hermann Bacher in the year 1989 in the drought-prone district of Ahmednagar. Ahmednagar is one of the largest districts situated in the central part of Maharashtra. It lies in the rain shadow belt with a hot and dry climate for most part of the year. The approach of IGWDP rested on community participation and self-help efforts organised around revitalisation of the local environment. The IGWDP was implemented by the formation of an organisation, Watershed Organisation Trust (WOTR) established in 1993 to undertake watershed developmental activities in resource fragile rainfed areas.

Watershed development marks a turning point that led to the conversion of the dry land to wet and irrigated land in the villages of Ahmednagar, Maharashtra. The villages of Ahmednagar display a radical change forwarding a march from poverty towards self-reliance through self-help. With more and more of area coming under wet land there was an alarming growth in farming economy, the cash crops and the dairy, besides other livelihood opportunities, through availability of more of agricultural labour for the landless reducing distress migration.

Facilitation and strengthening of community institutions is a part of integrated and participatory watershed development. With the increase in agricultural output and the women self-help groups, an additional development activity to be taken up with the participation of women in the villages was thought of. Participatory growth monitoring of children 0–5 years was taken up as an independent activity by Sampada Trust, sister organisation of WOTR and an NGO in Ahmednagar addressing women's felt needs and child health. The growth monitoring of children initially started in watershed villages is undertaken in areas even with no watershed development. Growth monitoring is a development input with child care components involved.

A study was conducted with a purpose of making an impact analysis of integrated watershed development in place for long and the new initiative of participatory growth monitoring of children 0–5 years in watershed villages. As the idea behind starting a development activity like participatory growth monitoring is linked with watershed development, it is considered as a part of watershed development. Hence, integrated watershed development and participatory growth monitoring together are referred as watershed development inputs in the book. With these development inputs triggering the processes of socio-economic and health change, a change in health consciousness was desired. A question that emerged was: "Does the raised income level through watershed development contribute to a rise in community nutrition/health consciousness and nutritional status of children?" Or, "does an additional activity like that of participatory growth monitoring of children 0–5 years needs to be introduced to bring about the desired health change?" These were the concerns of the organisation having lifelong grassroots level experience in watershed development. A large-scale social science research study was designed addressing these concerns.

Documentation Pathways

The large-scale study so designed has two components: Quantitative study focusing on examining the nutritional status of children 0–5 years in relation to watershed development inputs. Qualitative study focused on understanding the nutrition and health consciousness, the health culture. Health culture refers to perceptions and practices related to child nutrition/health pertaining to children 0–5 years. The qualitative study served as a cross-check against data collected from quantitative study and as a method to have an insight into the process of change and rural development.

The book discusses results of qualitative study in detail. The study makes a comparison at three levels. The qualitative study is conducted in three villages in Ahmednagar, namely Mhaswandi, Kangar and Kalas Pimpri. The villages belong to three different comparative study groups. Mhaswandi and Kangar are the villages where integrated watershed development has been implemented. Mhaswandi differs from Kangar in having additional activity of growth monitoring of children 0–5 years being conducted. Kalas Pimpri is a village having no inputs of watershed development and growth monitoring. Except the difference in development inputs, the villages are similar in cultural and environmental characteristics. The ethnic composition and the mode of living are similar. There are two major ethnic groups: the Hindu caste Maratha and the tribal communities, viz., the Thakar, the Bhil and the Lamani found in three villages.

The book discusses insights obtained in the processes of change and rural development in semi-arid regions as facilitated by watershed development inputs. Development processes across communities and women folk are discussed at length. Discussions surrounding knowledge and beliefs concerning factors affecting child nutrition and health, feeding and dietary practices of children form the main text. The nutritional status of children as per *Anganwadi* records is presented in the light of nutrition culture change. The linkage of watershed development inputs and change in the culture of child nutrition is discussed with a holistic perspective.

The data have been collected by means of informal in-depth interviews with elderly people, focus group discussions with women, mothers and care givers, key informant interviews, informal conversations and observations. While the mode of data collection for these tools was semi-structured, the data on village profiles have been collected using a structured format. Researchers with a background of anthropology with assistance by social workers and students of Rural Chroniclers Course of WOTR and educated unemployed rural youth formed the study team. Prior to data collection, introductory meetings were held in all the villages to discuss about the study, the team members, stay in the villages and support was requested. Consent for the study was obtained verbally at village level during the meetings and from the informants participating in the interviews. An oral consent form having a written script was developed in local language that detailed out the topic of the study and the work to be carried out. The researchers received a warm welcome and support for the study all over.

Social and Health Culture Change

Watershed development has contributed to significant economic growth as visualised from the increased crop yield and dairy development. Inter- and intra-community relationships are reinforced. Where post-watershed development contacts and inputs of NGO are continued, the women exhibit a great change in terms of exposure, mobility, participation in village development processes, decision-making, confidence and this reflects in raised consciousness pertaining to child nutrition and health.

A Concept Points Scale measuring culture change has been developed that displays change in health perceptions of three groups, the caste Maratha, the tribal community and the *Anganwadi* workers. Village Mhaswandi with watershed development and participatory growth monitoring inputs shows accelerated change on the scale. It is found that in watershed areas, an additional developmental input of growth monitoring of children 0–5 years, women institutions, collaboration between NGO and government workers and community participation are the key components that have a significant role in the enhancement of child nutrition and health.

The development processes that have set in, show a different picture when the data are viewed in terms of sex ratio of children 0–5 years and the girl child mortality figures. The prosperity that has occurred due to watershed development seems to permit expenditure on sex selection in dominant caste community. There is a change from surplus to commercial economy among the caste and change from subsistence to surplus economy in tribal community. Onion production and the rise in dairies often associated with watershed development leads to huge profits for the caste Maratha. The caste Maratha are heading towards commercialisation. Patriarchal culture with a value for men who have a significant role in agriculture, commercialisation, easy cash at disposal in dominant caste community are the social features at background behind the declining sex ratio. Gender discrimination exists before birth in the watershed village with growth inputs and after birth in watershed village with no growth inputs. The book discusses these social phenomena at length. More studies are needed to identify the watershed development and cultural factors triggering change in gender relationships as evident from skewed sex ratio.

There is a change from subsistence to surplus economy among the tribal community. The recognised expertise of the tribal farmers in tomato cultivation, rising health consciousness, increasing self-help groups of tribal women are notable changes in a community where the features like isolation, less exposure to the outside world, subsistence economy, alienation from market forces, from credit and saving and deprivation are widely known. A change in these conditions is reflected in enhanced child nutrition and health consciousness of community. A rise in the level of awareness in the child nutrition and health component helps in better management of the nutritional problems of the children for tribal women. More studies are needed to understand change associated with watershed development inputs among the tribal communities and the status of the tribal communities in the changed scenario.

The strength of women institutions in watershed development and growth inputs group is remarkable in the development processes of the villages. All these forces result into greater awareness of child nutrition and health, raising the nutritional status of children 0–5 years.

In the absence of watershed development inputs attributes in women like greater degree of confidence, exposure and awareness in a small community are probably due to cultural variation. These attributes are not reflected in the health consciousness of the community as found in watershed development and growth inputs group.

Policy Considerations

Development inputs designed for income generation or for socio-economic growth must be associated with health inputs. Nutrition and growth inputs for children of age 0–5 years could be a priority considering the grave problem of malnutrition in India.

Concentration on development efforts raising awareness in the following areas would make the development and health input programme in rainfed and other rural areas more effective:

- Significance of birth weight in the nutritional and health status of the child.
- Significance and facilitating promotion of colostrum feeding.
- Weaning: Appropriate age of weaning in rural areas, importance of weaning food in the status of child nutrition and health.
- Appropriate frequency of supplementary food.
- Maintenance of record of immunisation of the children migrating out with parents.

Gender equity remains a grey area where development efforts concentrate on economic growth and prosperity. Declining sex ratio in a prosperous village with watershed development and growth inputs in addition is an example. The development efforts must concentrate on building consciousness of the community in valuing girl child right from planning and implementation of the programme. The role of women institutions would be primordial in these efforts.

One of the important findings is that the dominant caste, due to its numerical strength, economic and political hold exercise control over natural resources in the interest of watershed development and have greater say in the decision-making processes. These approaches cause disadvantage to the tribal community. Support and inputs from watershed development organisations maintaining the interests of the tribal livelihood are pertinent in this matter. Community sensitivity of the development staff trained in value neutrality is the utmost need. Development inputs are essential in the following areas:

1. Raising negotiation skills of the tribal community so that they can voice their concerns in a better manner.
2. Raising consciousness of caste community regarding values of village development as a holistic concept, importance of the interests of the livelihood of the poor, values of social and gender equity are essential without which development is not meaningful.

Watershed development efforts with participatory growth monitoring inputs initiated yield better results in terms of child nutrition consciousness and status. Watershed development is a technical intervention having social implications. Social/community equity, gender equity, nutritional status are some of the important parameters which watershed development needs to take into account right from the processes of planning so as to make development more meaningful and sustainable. The book discusses these issues in detail. Policy implications are highlighted in the chapters.

Research Considerations

More studies are recommended to validate the results of the current study. Cross-cultural studies and studies from different geographical regions would bring out additional watershed factors contributing to change in health culture to the fore. Studies from the point of view of health culture change are meagre and this study is just a starting point for further exploration on similar lines. Watershed development has not been adequately explored from social science perspective. There is a dearth of studies on watershed development as triggering social, cultural, health changes, livelihood of the poor and disadvantaged, gender relationships in rural and tribal communities exploring facilitating and deterring watershed factors. Such studies will contribute not only to development policies and grassroots work but also to theoretical and methodological advancement in the social science academics.

1

Watershed Development Inputs and Child Nutrition

Watershed Development

Livelihood security and environment sustainability are inherent in the agenda of development policy and programme. Watershed development assumes great significance in this context specifically in the areas of rainfed agriculture characterised by low productivity and degraded natural resources gripped with poverty.

Watershed development is a technical intervention that works by capturing scarce rainwater and improving the management of soil and vegetation. Rainwater is trapped and harvested with ridge to valley bases in order to make water available year round to all the people living in a particular area. Watershed development is important for the overall development of the village. It is considered as the core development input for poverty alleviation in semi-arid and rainfed areas.

Greater food and livelihood security are ensured due to increase in land capability, increase in agricultural yield and allied income opportunities and by arresting the process of distress migration.

Maharashtra

A large part of Maharashtra falls under the semi-arid climatic zone and a large part of it is prone to droughts (WOTR, 2005). Maharashtra has faced recurrent droughts and famines several times in short duration periods from 1907 to 1920 after which the drought occurred in 1972. Agriculture being the primary means of livelihood of the rural poor in drought-affected areas, greater attention was focused on agricultural improvements in non-irrigated areas (Samuel et al., 2007). Obviously watershed development was looked upon as a means to bring about the desired change.

Watershed development has a long history in Maharashtra. In 1981, the Government of Maharashtra launched the Comprehensive Watershed Development Programme (COWDEP). During 1982–86, independent projects on watershed development were initiated by non-government organisations (NGOs) in Ahmednagar district. For example, work in village Kasare by an NGO, Social Centre from Ahmednagar and village Ralegan Siddhi by a social worker, Anna Hazare (WOTR, 2005).

Samuel et al. (2007) have listed out the large number of programmes that are in existence in the state such as the following. These programmes indicate the efforts undertaken for better productivity of land in rainfed areas since long:

1. Indo-German Watershed Development Programme, which is a bilateral programme.
2. Projects implemented by non-government organisations (NGOs) with support from local and foreign sources.
3. National Watershed Development Programme for Rainfed Areas (NWDPRA).
4. Drought Prone Areas Programme (DPAP).
5. River Valley Projects (RVP).
6. Integrated Wasteland Development Programme (IWDP).
7. Western Ghat Development Programme (WGDP).
8. Integrated Watershed Development Projects (IWDP).
9. Adarsh Gaon Yojana (AGY).

Within a span of ten years, 1980 to 1990, watershed development got rooted as an appropriate development programme for rainfed rural areas.

Ahmednagar, the largest district situated in the central part of Maharashtra has a fair background of watershed development. It is a rain shadow and drought-prone region. Ahmednagar is situated between 18.2 to 19.9 degrees north latitude and 73.9 to 75.5 degrees east longitude in the state. The total area of the district is 17,41,271 hectares, out of which 4,25,100 hectares have been brought under irrigation and 1,33,356 hectares remains non-cultivated (http://ahmednagar.gov.in).

Although a very small part of the district is irrigated, the villages in drought-prone Ahmednagar witnessed the process of change from poverty towards self-sufficiency. One of the major programmes, which caused the transformation of villages in Ahmednagar is the Indo-German Watershed Development Programme (IGWDP) initiated by Father Hermann Bacher in the year 1989. Father Bacher is the founder of Social Centre, the NGO in Ahmednagar.

The IGWDP's approach to watershed development rests on a community-driven participatory initiative which demands direct involvement and ownership stakes of the local inhabitants concerned. The transformation of several villages like Adgaon, Kasare, Pimpalgaon Wagha, Mendhwan from Ahmednagar district, are notable examples of success of watershed development works undertaken by IGWDP. Village Ralegan Siddhi, Ahmednagar set up a remarkable example proving the strengths of watershed development. These events led to a rise of watershed revolution in the semi-arid state of Maharashtra (WOTR, 2005).

Indo-German Watershed Development Programme

The Indo-German Watershed Development Programme (IGWDP) is a multi-sectoral programme. The programme involves integration of components of development such as:

- Human Resource Development (Community Development)
- Soil and Land Management

- Water Management
- Agriculture Development
- Livestock Management
- Afforestation
- Pasture Development

In the IGWDP, the area of the watershed is generally between 500 to 1,500 hectares. The population ranges from 750 to 2,000 (WOTR, 1997).

The success of watershed development depends on community participation. Community participation is the core principle of IGWDP. The community plays an active role right from design of the project, its implementation and maintenance.

Building and strengthening of people's institutions is an important part associated with IGWDP. These institutions form the means through which various development works can be initiated, viz., formation and sustenance of Village Watershed Committees (VWCs) now called as Village Development Committees (VDCs), women's committees like the Self-Help Groups (SHGs), the *Samyukta Mahila Samiti* (SMS), the apex body of SHGs and others. These institutions ensure community participation with greater involvement of women who are the key to successful development.

Watershed development under the IGWDP involves a multidisciplinary approach that involves continual interaction and exchange between and amongst the various sectors and disciplines. The programme is comprehensive, integrated and participatory (WOTR, 2007).

The concept of watershed development as used in the present refers to the approach based on IGWDP.

Resource Organisation in Watershed Development in Ahmednagar

The IGWDP was implemented by the formation of an organisation, Watershed Organisation Trust (WOTR) established in 1993 to undertake developmental activities in resource fragile rainfed areas. WOTR is a non-government official capacity building organisation of the Indo-German Watershed Development Programme.

WOTR now a resource organisation was known as mother NGO working in watershed development in Ahmednagar district in Maharashtra for a period of five-six years, during 2002–2007. The concept of mother NGO was introduced by the Government of Maharashtra in the year 2001. Many different NGOs and community-based organisations, village groups have carried out the work of watershed development with the technical support of WOTR.

The main focus areas of WOTR are capacity building of village groups and NGOs for participatory watershed development, self-help promotion, direct implementation of watershed projects, micro-credit activities, training and extension for organisations and practitioners, development of concepts, pedagogies, training manuals, awareness generation tools and media aids, documentation and action research.

Participatory Growth Monitoring of Children 0–5 Years

The field experience of WOTR indicated that with more availability of cash from increased crop and milk yield as triggered by watershed development, health expenditures were simultaneously on the rise. Whatever the community earned through watershed development spent a large amount on treatment for illnesses. When the reasons for taking loans by the women from Self-Help Groups (SHGs) were analysed by the WOTR field officials, they noticed that in most cases the money was taken for the treatment of small children. A need to draw the attention of the community towards enhancement of their own nutritional and health status rather than spend it on the treatment of illnesses was realised. It was felt that inputs related to child nutrition and growth would increase the effectiveness of overall village development. This led to the idea of facilitating the process of change towards child development with a focus on improving the nutritional status.

As health of children assumes significance for human and economic development and most of the health problems among children, specifically in the age group 0–5 years are preventable and are mainly related to poor nutritional status, a need to undertake a developmental activity of participatory growth monitoring of children 0–5 years was put forth by WOTR. A full-fledged programme on child care and growth monitoring emerged. The project has been undertaken by Sampada Trust, sister organisation of WOTR, and an NGO in Ahmednagar. The main focus of work of Sampada Trust is on women empowerment. The work on growth monitoring started with the watershed villages in the year 2005 (Sampada Trust, 2005 and 2006).

The Participatory Growth Monitoring Programme, as it is called, includes the following components:

1. Community awareness generation regarding nutritional status of children.
2. Promotion of women health promoters *(Mahila Pravartak)*.
3. Regular height and weight check up of children 0–5 years.
4. Workshop for *Anganwadi* workers and women health promoters.
5. Facilitating collaboration between *Aanganwadi*, health promoters, SHG women and community.
6. Demonstration on balanced diet.
7. Children and parents meet.
8. Healthy baby competition.

The activities numbered 6 to 8 are carried out once in project duration. The project duration ranges from six months to one year. The activities like undertaking measurement of height and weight by health promoters continues further. Therefore, the key feature of the programme is monthly monitoring of weight and height of children 0–5 years. The weight is taken on spring balance weighing machines and height is measured using a tape provided by Sampada Trust. The health promoters pay home visits and record the measurements. They inform the parents then and there about the nutritional status of children, guide the parents of low weight children to provide

Plate 1

A Mother Plotting Weight of her Child on Growth Chart

nutritious diet to children. For example, feeding children with food items like sweet *sheera/rawa* using jaggery and adding fruit like papaya grown around the houses, other items like *bhaji/pakoda* (Fritters: a piece of food item coated with batter and deep-fried and is spicy) using gram flour and adding soyabean nuts. Supplementing food with items like groundnut and jaggery, fruits like bananas rather than *bhel* to eat regularly. Children like *bhajis/pakodas*. Gram is grown in the field and use of soyabean oil is popular in the village. So people know about soyabean and they like it. Another important feature is that the mothers of children are involved in plotting the weight of their children on the growth charts developed by the organisation. It alerts them on the nutritional status of their own children.

Health promoters are women promoted by the community and who have voluntarily taken up the task of monitoring the weight and height of children 0–5 years. They are guided and trained by Sampada Trust to take care of the varied components of the child care and growth monitoring project at village level.

Sampada Trust is involved in a number of activities. Some of the important activities are:

1. Promotion of Self-Help Groups (SHGs). These SHGs are federated in an apex body at the village level called as *Samyutka Mahila Samiti* (SMS). Two members from each of the SHGs are represented in the SMS.
2. Capacity building of village women through trainings for management of SHGs, entrepreneurship, child care and growth monitoring of children 0–5 years.
3. It is through the SMS that micro-credit and technical information is provided to SHG women for entrepreneurial activities and provision of micro-insurance.

The activity of growth monitoring of children is not a new initiative. Government ICDS (Integrated Child Development Services) programme is being implemented through *Anganwadis* in the villages for long. The concept of growth monitoring as defined here is understood in relation to watershed development. It refers to the Sampada Trust, an NGO-initiated activity with the above-mentioned components. The introduction of components of growth and nutrition in watershed villages helps to enhance the effectiveness of programme and village development as a whole. The activity addresses women's felt needs; and the health of children those can make headway towards the sustainability of watershed development processes.

Development inputs as mentioned in the text therefore refer to two types of inputs:

1. watershed development with the components as mentioned, and
2. NGO-initiated participatory growth monitoring of children 0–5 years with the components as mentioned.

Child Nutrition

Nutrition is of prime importance in the attainment of normal growth and development and in the maintenance of health throughout life.

Adequate nutrition is critical to child development. Undernutrition is the pathological state resulting from insufficient or less quantity and quality of food (Jelliffe, 1966). 'Malnutrition' is the term often used to point towards undernutrition. However, malnutrition includes both under and overnutrition. In the present discussion, the studies mentioning about malnutrition refer to the state of under nutrition.

Food and Social Processes

Pelto and Freake (2003) point out that in social science research, nutrition issues are observed from a wide variety of disciplines. They point out that food and nutrition issues are vehicles for examining social processes.

Social change is one such process, which can be observed on the basis of nutrition issues. The complex social and cultural factors, which mediate interactions leading to particular nutrition outcomes, would speak of the process of social change that occurred. Impact assessment of watershed development inputs provides scope and an opportunity to understand the interactions of development and culture resulting in change as observed in terms of child nutrition and health.

The Problem of Malnutrition

Deshprabhu (2001) mentions malnutrition as the leading killer in the world, killing 5–8 million children each year, accounting for 10 per cent of all deaths.

South Asia has the highest prevalence of malnutrition in children. It is higher than even sub-Saharan Africa which in many ways is lesser developed than South Asia (Bamji, 2003).

Nearly half of all malnourished children in the world reside in the few South Asian countries, India being the largest (Bamji, 2003; Ramalingaswamy et al., 1996).

A report by *Indian Express* (3 May, 2006 online edition) highlights a UNICEF (2006) finding that of the 146 million children, under five, are undernourished in the developing countries and India alone harbors 57 million (39%).

National Family Health Survey (NFHS-3) highlights the problem of child malnutrition in India. Almost half of children under age five in India are stunted (short for age), 20 per cent are wasted (thin for height) and 43 per cent are underweight, which takes into account both chronic and acute under nutrition. During the first six months of life, 20–30 per cent children are undernourished. There are significant sex differences with girls reporting lower values. Children in rural areas are more likely to be suffering from chronic under nutrition (International Institute for Population Sciences and Macro International, 2007).

NFHS report indicates child nutrition and health remains a major challenge to overcome in the country. Inadequate nutrition is pointed out as one of the major factors responsible. This could be because of either inadequate availability of food or because of inadequate feeding of the child due to ignorance. Watershed development can play a major role in impacting upon food availability through increased food production while participatory growth monitoring would be significant in dealing with the ignorance. A blend of these two inputs may yield fruitful results.

Causative Factors

Inadequate food/nutrient intake and diseases affecting the nutritional intake or metabolism have been identified as major contributory factors of malnutrition. Poor child feeding practices largely explain the low nutrient intake (Kapur et al., 2005; Kulkarni, 2008). Malnutrition is caused by a combination of factors. A shortage of food as well as proper utilisation of food are identified as causing malnutrition.

Insufficient feeding in ignorance even when adequate supplies are available is also a problem. Late weaning with poor food is the leading child health problem in developing countries. When weaned, many children are given adult foods they cannot chew or digest, or that are unnourishing (Deshprabhu, 2001).

There are many biological and cultural aspects which influence the nutritional status of a person, for example, the age and sex composition, culturally controlled food preferences and restrictions, education, occupation, general health condition, etc. (Kulkarni and Sarma, 2001).

Malnutrition is a complex web with a wide number of causative factors. A free list of causative factors leading to under nutrition has been prepared and presented in diagrammatic form (see Figure 1). The list is based on the studies done in the recent times as follows:

Free list of factors affecting under nutrition among children based on review of literature:

1. Poverty
2. Food shortage
3. Poor sanitation
4. Housing conditions
5. Inadequate access to safe drinking water
6. Personal hygiene
7. Low sex ratio
8. Women's participation in environmental regeneration
9. Women's education
10. Maternal malnutrition, pregnancy care
11. Maternal mortality
12. Women's age at marriage
13. Age at first childbirth
14. Family planning
15. Infectious diseases: diarrhoea, malaria
16. Low birth weight
17. Child's age and sex
18. Birth intervals
19. Failure to breast feed
20. Poor and late weaning practices
21. Poor feeding practices

22. Low intake of diet
23. Immunisation
24. Access to health care services
25. Belief system
26. Failure to use food properly

(source for free list: Bamji, 2003; Brennan et al., 2004; Das, 2008; Deshprabhu, 2001; D'souza, 2001; Kapur et al., 2005; Khadilkar et al., 2007; Kulkarni and Sarma, 2001; Venkateswarlu and Reddy, 2000; Yadav and Roy, 2005).

Role of Watershed Development

Poor land productivity and resultant poverty are linked to malnutrition in children. The role of watershed development is critical in reducing malnutrition as it ensures food and livelihood security. A study by Kirsten et al. (1998) in South Africa highlights agricultural activities make a positive contribution to household nutrition. Designing effective programmes for improving agricultural productivity could have a potentially positive impact on household and child nutritional status.

Food availability is greatly affected by its production. Watershed development has the potential to bring about the desired change in the factors impacting upon the nutritional status of rural children, of which food production is the major.

Watershed development as a means to alleviate poverty is being extensively implemented across the country to drought proof degraded lands. The land productivity as measured by crop production is greatly enhanced (505%) and households owning cross-bred milch cows also increase (41.2%) as compared to the pre-watershed situation (D'souza, 2001).

The changes in crop and milk production with availability of water leave a profound impact on food and nutritional status of the population.

Impacts of Watershed Development

The impacts of watershed development have been widely recognised. During the initiation of watershed development programme, facilitation, formation and sustenance of people's institutions, viz., Village Development Committee (VDC), women's institutions, the SHGs and the SMS form an important part of integrated watershed development programme. The approach of integration of different components of development with participatory management of the programme makes a significant impact leading to overall holistic and sustainable development of the village. They have shown the potential of doubling the productivity of rainfed areas while sustaining the natural resource base.

Records show that the effects of the drought have been greatly mitigated wherever watershed activities have been undertaken in a comprehensive manner with participatory efforts, that is, by the people themselves. The villages no more need tanker services for drinking water and migration is less common than during drought conditions.

Studies highlight the impacts in terms of the alarming rise in average income of the villages and households. Recent articles in a leading newspaper and a reputed science magazine have recorded the impacts of watershed development in transformation of villages. Enhanced agricultural productivity and milk production contributing to increase in average income are mentioned several times in the studies and the recent articles. Hiware Bazar, a village in Ahmednagar district and Gawdewadi, a village in Pune district in Maharashtra have attracted the attention of the media.

An alarming growth was taken note of in the following components in the treated watersheds:

1. Increase in groundwater availability, e.g., increase in the number of wells in the villages.
2. Increase in biomass.
3. Increase in land productivity and milch animals with consequent rise in milk production.

4. Rise in household, village and cash crop income.

The other changes associated with watershed development are as follows:

5. Greater participation of women in village development processes.
6. Greater gender balance in the family and community through their involvement in watershed at every stage.
7. Increase in awareness regarding watershed, environment, sustainable agriculture, community's own traditional knowledge.

In many villages of Ahmednagar, before the implementation of watershed there was not enough fodder for animals. With watershed development, villages have planted trees. Watershed development work is implemented on three conditions in the villages. They are ban on free grazing, ban on tree cutting and contribution of voluntary work. Because of the bans there is conservation of trees and shrubs. Farmers cut the grass and use it as animal fodder. Because of this farmers have purchased cross-bred cows yielding more milk. Dairies have been started in some of the villages giving higher yields.

The rise in employment opportunities in agriculture and dairy has led to a rise in income. More cash than before is available. It is assumed that this has led to the increased intake of food and milk at the household level consequently enhancing the nutritional status of children (*Watershed Voices*, Newsletter on Mhaswandi Watershed Project).

While a reduction is noted in the following components as exemplified in the literature:

1. *Poverty:* For example in Hiware Bazar, there is 73 per cent reduction in poverty, due to profits from dairying and cash crops. Gawdewadi, a village, in Ambegaon taluka was transformed into a flourishing agro and milk produce centre and has received several state level awards. These villages have become major centres of milk collection.

2. *Migration:* For example, people who migrated in search of work have returned to their home villages and are self-reliant now.

(sources: *Down to Earth*, 2008 and *The Times of India*, 2008 and WOTR, 2005).

Watershed Development—Growth Inputs and Child Nutrition: What are the Links?

Watershed development helps enhance food security is observed from the studies and field experience. To what extent watershed development impacts on nutritional security and enhancing the nutritional status of the population would speak of its effectiveness, sustainability and the direction of change it has led.

Shah (2004) highlights that a number of studies have been conducted over the past decade assessing the impact of various watershed projects in the country. These studies are conducted from the point of view of monitoring and evaluation with a purpose to report to funding agencies both in the government and the non-government sectors. The impacts are primarily measured in terms of economic benefits and environment regeneration.

The studies indicate watershed development has not been truly assessed in terms of nutrition issues. There are quite a few studies related to understanding of the issues in malnutrition and finding out the various causative factors. The role of development inputs impacting on nutrition have not much been a focus of interest. Watershed development being a critical input for food security, it would be meaningful to understand its impact on nutrition related phenomena, the nutrition culture and the status of children of age 0–5 years in particular.

A study has been taken up to understand the role of watershed development in contributing to change from the point of view of child nutrition conducted at WOTR. Participatory growth monitoring (PGM) of children 0–5 years has been undertaken by Sampada Trust in the villages of Ahmednagar as an independent activity. As the activity is conducted in watershed villages too, it was thought that understanding the role of PGM in child nutrition in comparison to watershed development would be worthwhile. Hence, the roles of two inputs, watershed development and participatory growth monitoring were taken up to understand the impacts on child nutrition in comparison. It was thought that the study would give a direction to watershed development programme as a whole. The present work concerns with this comparative impact study.

A framework enlisting factors affecting child nutrition has been developed based on the conglomeration of available literature and field experience in watershed development. The framework is presented in diagrammatic form as an aid to researchers working in the area of watershed development. A note explaining the diagram in detail is given at the end of the diagram.

Figure 1

Diagram showing the factors affecting child nutrition and health and the role of watershed development and growth monitoring inputs

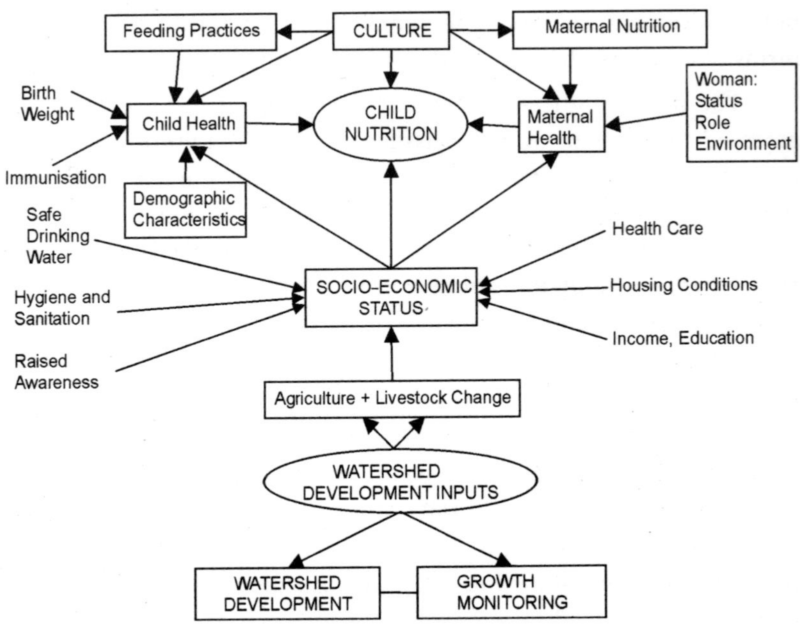

The diagram shows that watershed development and participatory growth monitoring referred as watershed development inputs have a role in changing the child nutrition and health culture. However, what factors of these development inputs cause the change and the kind of change they bring about are not exactly known. Exploration of these factors and the path of change: social, cultural, health is the subject matter of the study on which the book is based.

Descriptive Note on Figure 1

Culture of a community and socio-economic status are the two major components that contribute to child nutrition. Besides, mother's health significantly contributes to child health and the intervening social factors in child nutrition and maternal health are interrelated.

Watershed development contributes to enhanced socio-economic status by means of agricultural productivity and livestock development. The socio-economic status is a broad concept and includes a wide array of factors, viz., status of income, education, housing conditions, access to health care, access to safe drinking water, personal hygiene, sanitation, raised level of awareness of development processes at local level, health awareness.

Income here means increase in agriculture yield, rise in livelihood opportunities, for example, availability of agriculture labour, other wage labour work leading to availability of more cash than before. Education means access to higher studies, higher secondary level schooling due to rise in income and employment opportunities. A change in housing condition includes changes like building permanent structure, allotting separate place for the animals, reducing cost on repairing of temporary structures. Personal hygiene is improved with more water available for body cleanliness as per community perceptions. More cash is available to access health care. There is improved awareness regarding health. Raised awareness leads to insistence on means to make water safe for drinking. There is improved awareness related to balanced diet enhancing child nutrition and health due to growth monitoring inputs.

Culture governs the practices of the community towards child health. Birth, feeding notions and concepts of health, status and role of women in the society, role of women in environment generation contributes to child nutrition and health.

Watershed development with initiation of participatory growth monitoring activity brings about a change in consciousness leading to health promoting culture. A social change with a health promoting culture is likely to lead towards sustainable development.

Discussion

Watershed development has critical implications for child nutrition but the linkages have not been adequately explored. Not much is known about the pathways of watershed development factors affecting child nutrition.

There are exclusive studies related to child nutrition and studies related to the impacts of watershed development that speak of enhanced food and other crop production but the impacts on nutrition consciousness are not adequately known. The interrelations of development and changing nutritional processes need to be explored from the point of view of understanding the effectiveness of the programme.

The present study while addressing development concerns also attempts to work on the existent gap in knowledge and explores the linkages between watershed development inputs (inputs of watershed development and those of growth monitoring together) and child nutrition with an idea to provide relevant inputs for watershed development programme and policy planning.

In the field of watershed development, child nutrition as a vehicle for examining social change has not been adequately attempted. Development interventions are the dominant vehicles for conscious social change (Reeler, 2007). The long-term impacts of watershed development in Maharashtra on the one hand and the introduction of participatory growth monitoring of children 0–5 years by an NGO in watershed villages on the other hand leaves a lot of scope to assess the impact of the two independent inputs in terms of the changes that have and are occurring in the area of child nutrition.

In social change studies, the area of Health has not been adequately explored. International Development Research Centre (IDRC) states health as one of the indicators to measure development and nutrition as an important determinant of health (Bamji, 2003). Improved levels of nutrition and childcare are the criteria of achievement of health of community. Two of the Millennium Development Goals are related to child nutrition:

1. One is the *eradication of hunger and poverty*. Achievement of the goal depends on agriculture development, the primary occupation in rural areas, and
2. *Reduction of child mortality*. Child mortality in rural areas is closely associated with malnutrition. Premature deaths are more common among malnourished children than those with normal weight (Braun, Swaminathan and Rosegrant, 2004).

As a part of watershed development input, change in the area of child nutrition and health would help in understanding the processes of social change and rural development in semi-arid regions.

The current study assesses the impact of different watershed development inputs as an integrated whole in terms of change at the level of community: the perceptions and practices, understanding benefits and limitations of the development inputs with emic (people's perspective) and etic (researcher/outsider's perspective) perspective and gender issues, all in the context of child nutrition.

A study was designed to understand the process of social and cultural change in the area of child nutrition as impact assessment of the two kinds of development inputs. The study documents the percolation of change at the level of perceptions and practices/ actions of the community. An understanding of the impacts of growth monitoring inputs would help in assessing the need for introducing child growth inputs in watershed development programme. Considering the long history of watershed development and the new initiative of participatory growth monitoring of children 0–5 years age group in Ahmednagar villages by Sampada Trust it was WOTR's own interest in understanding the independent impacts of the two inputs. It was decided to study the impacts in terms of nutritional status of children 0–5 years and in terms of community consciousness of nutrition and health of children 0–5 years. Hence, a large-scale study having a blend of quantitative and qualitative components was designed. The present work discusses the results of qualitative study. As there are two types of development inputs whose impact was to be studied, it was decided to make a comparison at three levels in the study. That is, understanding the long-term impact of watershed development at one level, with watershed development where participatory growth monitoring has been initiated at another level with no such type of inputs given at the third level.

Roche (1999), stating the experiences of Oxfam and Novib, says "impact assessment is the systematic analysis of the lasting or significant changes—positive or negative, intended or not—in people's lives brought about by a given action or series of actions".

The process of social change traced here is an impact assessment of watershed development inputs. There is a chance that more weightage is assigned to watershed development inputs than other factors in the analysis of change. The objective was to assess change in the areas of inputs and no inputs. This does not undermine the role of factors like education and others in the process of change and development.

The movement of watershed development has taken deep roots in the semi-arid district of Ahmednagar, Maharashtra. The initiative of participatory growth monitoring introduced in watershed providing inputs for enhancement of nutritional status of children by an NGO is in place for about two years.

If health is to be included as a component of integrated watershed development as a necessity for sustainability, it is important to understand the independent impacts of the two development inputs.

2

Pathways of Understanding Change

Understanding Social and Cultural Change

For impact assessment it is necessary to understand the time period after which the impact is visualised. The watershed work under IGWDP is completed in four to five years and the impact is visible in all parts of the treated watershed after completion of five years of implementation (*Watershed Voices*, Newsletter on Mhaswandi Watershed Project). The impacts of participatory growth monitoring are not yet fully understood.

A large-scale comparative study designed to understand the impacts of watershed development and participatory growth monitoring involves two major components:

1. Quantitative study
2. Qualitative study

Quantitative Study

The quantitative study concentrates on assessment of nutritional status of children 0–5 years by means of anthropometry and method of 24 hours diet recall.

It involved a detailed village level and household level survey, data collection on children's schedule: feeding and food consumption pattern and mother's profile. The quantitative study was designed to understand watershed development related changes in socio-economic status of the household, the women and the children 0–5 years.

A clinical examination of children 0–5 years was undertaken in selected few villages as an additional and side activity for assessment of nutritional status of children with a clinical perspective.

Qualitative Study

The qualitative study was conducted to serve two important purposes:

1. One was to triangulate and supplement the data collected in the quantitative study and
2. Second was to have an insight into the background factors, the watershed development and the culture of which child nutrition and health is a part.

It is believed that the qualitative study would help strengthen the correlations and the generalisations made on the basis of quantitative study. The analysis and writings of the qualitative study were complete much before the results of quantitative study could be out. The empirical richness of the qualitative study permitted having a small book published.

The qualitative study focuses on understanding the community perceptions and practices related to nutrition of children 0–5 years. Although related, each study, quantitative and qualitative, is independent and vast. The present work discusses the results of qualitative study in detail.

Nutrition and health are aspects of culture. The status of a society's nutrition and health are governed by the perceptions held by the community. In any particular community, health-related knowledge, beliefs and practices are passed on from one generation to another through traditional education. Nutritional status of children is also a product of the thoughts and actions undertaken by a community. Community perceptions provide a cultural context and reasoning for the existent nutrition status of the children. The motto of the qualitative study was to understand the percolation of change as an impact of development inputs at the level of culture. A change is likely to become sustainable when it becomes a part of culture.

The study is based on the following assumptions:

1. That participatory growth monitoring is an essential input for the success of watershed development when observed in terms of nutrition and health consciousness of the community.
2. Watershed development alone is not sufficient input to bring about a change in child nutrition perceptions and practices.

Research questions:

1. Whether watershed development inputs (inputs of watershed development coupled with growth monitoring) as a whole has brought about a change in the thoughts and perceptions of the community and in behaviour, that is at practices level, with regard to the child nutrition?
2. What are they?
3. How have the changes come about?
4. Are they comparable?

Aims and Objectives of the Study

The aim of the study was to make a comparative analysis of the process of change as a part of development inputs on community perceptions and practices in the area of child nutrition with particular focus on age group 0–5 years.

The objectives of the study were to understand the following:

- Social changes influenced by watershed development and growth monitoring.
- Knowledge and beliefs affecting child nutrition and health.
- Feeding and dietary practices of children.

- Explore linkage between watershed development inputs and change in the culture of child nutrition.

Comparative Studied Groups

The study makes a comparison at three levels. It is conducted at three sites with different inputs and no inputs as follows: a comparison of one development input with another and the two inputs with a no input group:

1. A village where watershed development has been implemented and growth monitoring activity undertaken.
2. A village where watershed development has been implemented but no growth monitoring activity undertaken.
3. A village where no watershed development and growth monitoring activities are undertaken. As the study is qualitative, three villages from Ahmednagar district have been selected to conduct the data collection in-depth.

Studied District: Ahmednagar, Maharashtra, India

There are fourteen *talukas* in the district. The study is conducted in the villages of three *talukas*, viz., Sangamner, Rahuri and Pathardi.

Climate

Ahmednagar stands 1,900 feet above sea level. The climate of the district is hot and dry. Hot summer and general dryness during major part of the year is the characteristic feature, except during south-west monsoon season. The air in the cold season which lasts from November to February too is dry. However, in the hilly western part of the district, the climate is slightly cool. Field experience indicates that the period from February end to the first week of June could be said as the hot season. It is followed by the south-west monsoon season which lasts till the end of September. October and November constitute the post-monsoon or the retreating south-west monsoon season. The western hilly region receives more rainfall, but as one moves towards the east, the amount of rainfall decreases.

Rainfall

The average annual rainfall in the district is 501.8 mm (19.76"). It is plentiful in the hilly parts of Sangamner and Rahuri, the study talukas. In the plains, the early rains are often scanty and the late rains capricious, so that droughts, especially in the Bhima basin, appear to form the rule. The average rainfall received by each of the study taluka as follows speaks of the climatic conditions of the site wherein the study was conducted.

Table 1

Rainfall in Studied Talukas

Study Taluka	*Average Rainfall (mm)*
Sangamner	416.6
Rahuri	455.0
Pathardi	580.4

Source: http://ahmednagar.gov.in

Land

All cultivable land in Ahmednagar comes under the two great heads of *jirayat* that is dry crop and *bagayat* that is watered. Dry crop lands are either Kharif that is sown with the early crops, or Rabbi that is sown with late crops. The Rabbi crops are sown in June and July and reaped at the end of August or in October or November. The late crops are sown in October or November and reaped in February and March. In hilly west of Akola near the Sahayadris, the early crops including the rice and the coarser hill grains are the most important. During the cold weather a little wheat, peas, gram and lentils are grown.

The three major soils are black, red and gray including white.

Temperature

The month of May is the hottest month and December is the coldest month. On few occasions thin film of fog is being observed in the early mornings of December and January.

Studied Tehsils/Talukas: Sangamner, Rahuri and Pathardi

Considering the physical set-up of Ahmednagar district, there are three physical divisions: Western Hilly Region, Central Plateau Region and the Region of Northern and Southern Plains.

1. Part of Sangamner taluka is included in Western Hilly Region. The hill ranges of Adula, Baleshwar and Harishchandragad lie in this region and various high peaks are found in the same region. A part of Sangamner is included in Central Plateau Region too.
2. Rahuri and Pathardi talukas are included in the region of Northern and Southern plains. This is the region of the Godavari and the Pravara river basins.

Rahuri is the central subdivision. The soil is deep black coloured and retains its moisture for a long time. It receives better rainfall than Sangamner which lies on the West and nearer to the Sahyadri. Early and late crops are grown in Rahuri in equal proportion. Late crops are grown in plains.

Sangamner is the northern subdivision. The soil in hilly region is of inferior character. Various streams are dry in hot season and water supply is not good. Sangamner has on the whole better lines of communication than most of the other sub-divisions.

Pathardi is the chief cattle market while at Bodhegaon, a village in Pathardi where the largest transactions in grain are effected. Both places are residents of wealthy merchants. About 2,000 looms are in Pathardi, principal manufacture is of various kinds of cotton clothes.

(source for information on study district and talukas: http://ahmednagar.gov.in and Gazetteer of the Bombay Presidency, 2003).

Studied Villages: Mhaswandi, Kangar and Kalas Pimpri

The selection of the study site is purposive. As it takes at least five years to harvest complete potential benefits of watershed development the villages have been so selected where five years of implementation of watershed development are complete. Villages Mhaswandi from Sangamner and Kangar from Rahuri taluka fulfil the selection criteria.

The villages where the study has been conducted, viz., Mhaswandi, Kangar and Kalas Pimpri, represent three different comparative study groups. Watershed development has been implemented in Mhaswandi and Kangar. Mhaswandi is a village where participatory growth monitoring of children 0–5 years has been initiated by Sampada Trust since June 2006. There has been no such growth monitoring activity undertaken in Kangar. Village Kalas Pimpri is one where no inputs of watershed development and participatory growth monitoring are undertaken.

The villages have similar cultural and environmental background and belong to one district, Ahmednagar. The comparative nature of the study of the impact of watershed development inputs demanded selection of the villages with similarity in many respects except for the development inputs. The villages are similar in ethnic composition with cohabitation of two different communities, the caste and the tribe. The details are given in Chapter 3, 'Villages: A Profile'. The caste Maratha is numerically dominant in all three villages. All the three villages have semi-arid environment characteristics. The comparison has been made across the watershed development inputs. One village in each of the above- mentioned comparative study group is selected as shown in the following table:

Table 2 gives a picture of the villages selected for the study.

Participatory growth monitoring is an ongoing activity.

Table 2

Description of Study Villages

	Comparative Group	*Village*	*Taluka*	*Watershed Implementing organisation*	*Year of Initiation and Completion of Watershed Development (Records of WOTR)*	*Year of Initiation of Participatory Growth Monitoring (Records of SAMPADA TRUST)*
1.	Watershed Development (WSD) and Growth Monitoring (GM) activity conducted (WSD+GM)	Mhaswandi	Sangamner	Sangamner Bhag Sahakari Sakhar Karkhana (SBSSK), Sangamner	1993–2000	June 2006
2.	Watershed Development. No Growth Monitoring conducted (WSD)	Kangar	Rahuri	Rahuri Sugar Factory, Rahuri	1994–2001	Not applicable (NA)
3.	No Input of Watershed Development and Growth Monitoring (No Input)	Kalas Pimpri	Pathardi	NA	NA	NA

Ethnic Composition

In the study villages, the Hindu caste and the tribal communities have substantial population. The caste community Maratha outnumbers all other communities. The Mahar is the other caste community which has a considerable population. They are found in two villages, Kangar and Kalas Pimpri in more numbers and are few in number in village Mhaswandi. In Kangar, they rank second in major population group following the Maratha, while in Kalas Pimpri, they are in small numbers. The population of the ethnic groups is given in the village profile. The caste community Maratha belongs to the higher rung in the traditional Hindu hierarchy and is open caste category as per constitutional classification.

The Mahar is a Scheduled Caste in Maharashtra (Government of Maharashtra, 2006).

As the caste community Maratha and the tribal communities are found in more numbers in all three villages, the two groups offer a better opportunity for comparison of data across the villages. Hence, a comparative analysis and interpretation of change takes into account the caste community Maratha and the tribal communities.

Among the tribal communities, there are three different tribes found in three villages: the Thakar, the Bhil and the Lamani. The Thakar and the Bhil are the Scheduled Tribes (ST), while the Lamani is the Denotified Tribe (DNT) (Government of Maharashtra, 2006).

Scheduled Tribes and DNT is a constitutional classification. However, the cultural characteristics of the ST and the DNT are slightly different. The wandering tradition of the DNT from generation to generation makes the difference. The ST lead a comparatively settled life.

The settlement patterns of caste communities, the Maratha and the Mahar are different and individual. The tribal settlements are far away from the caste settlements. The tribal people reside on the hilly terrain while the caste communities on the plains.

Data Collection

Methods

The research work has been carried out in the rural parts of semi-arid and drought-prone Ahmednagar district of Maharashtra in India.

The study discusses the perceptions and practices of the communities in the area of child nutrition resident in comparative study groups, the three villages. The analysis and discussion pertains to the children of age 0–5 years.

Qualitative methods have been employed to collect the data in three villages. The importance of qualitative methods has long been recognised in development research. They have been found to be more appropriate in terms of providing insights for the development of culturally relevant programmes (Aube and Mansour, 1989).

The understanding of the communities regarding child nutrition and the affecting factors (facilitating and deterring) with reference to age group 0–5 years have been explored in the perceptions and practices.

The following tools and techniques have been employed for data collection:

1. Focus Group Discussions (FGD) with women.
2. In-depth Interviews (IDI) with elderly women and men and key informants.
3. Informal Conversations (IC) with people, individuals and groups and
4. Observations (OB).

Two different types of interview formats were developed: Structured and Semi-structured. The methods, the formats developed and the background of the researcher are described in the following table to understand at a glance.

The data on nutritional status of children was obtained from the *Anganwadi* centres of the three villages/study groups. The purpose was to find out if the results of qualitative data show any consistency with the nutritional status of children. The data on the nutritional status of children as recorded in *Anganwadi* are based on weight-for-age criteria. The weight of the children is taken using a spring balance weighing machine in *Anganwadi*.

Procedure

The data collection was done during the period from December 2007 to May 2008.

A structured format was developed to collect data on village profiles, the infrastructure and services available in the villages and data from *Anganwadi* on nutritional status of children. The list of informants with whom the interviews for village profiles were held is given in the table. The data on village profiles were collected by the students of Rural Chroniclers. 'Rural Chroniclers' is a course conducted by WOTR for the unemployed educated rural youth. An intensive training about the project and data collection was imparted to the students before they entered the villages.

A semi-structured interview and observation guideline was prepared to assist in collecting the data on community perceptions and practices. The in-depth interviews, focus group discussions and observations were conducted by anthropologists, the principal investigator and associate. Field assistance in fixing up the FGD and IDI, gathering the informants for the same and conducting small individual interviews and observations, was done by a team member, a social worker by training.

The focus group discussions (FGDs) were held with women. The women included primarily the mothers of children 0–5 years and/or caregivers like grandmother in case when mother was away due to migration for employment outside the village. One FGD was conducted per community. The FGD participants numbered from five to eight. The IDIs were taken with different people as listed out in the table. IDIs were also taken with *Anganwadi* teachers. IDI with officials of WOTR and Sampada Trust were taken to have their perspective on the changes that occurred in the villages. The informants for FGD and IDI were selected. Persons who did not have a chance to participate in the quantitative study, that is in survey, due to restricted sample size were involved in qualitative study. This was done to encompass the views and perceptions of a larger section of the village.

SHG formation is an important part of the implementation of integrated watershed development programme. In Mhaswandi, there are 23 SHGs. Most women from the village are members of SHGs. Village Kangar too has twelve women SHGs. The women members of SHGs who had children of age 0–5 years were involved in focus group discussions to understand their perceptions and practices related to child nutrition and health.

The interviews and discussions were conducted with consent and by fixing up time with the informants. Preference was given to their free time and convenience. The study team stayed in the village in the house of the villagers for data collection. The focus group discussions and interviews were held either early morning or late evenings after the people returned from the fields and had their dinner.

The details of the data collection procedure are summarised in the Table 3.

Photography as a tool supplementing the data has been used as and when an event was felt as important and related to the study.

Rapport Establishment

WOTR has good contacts and rapport in the villages where watershed development has been implemented. The rapport was further strengthened in the villages by developing relations from the point of view of the study. In no-input village, the rapport was newly established. Introductory village meetings were conducted before the data collection. This proved useful in developing initial contacts among the people. During the data collection, the researchers stayed in the villages. Accommodation and food arrangements were made in the houses of some key persons and who voluntarily made arrangements like a health promoter *(Mahila Pravartak)*, *Anganwadi* worker, local small hotel owner's house.

A researcher's stay in the village served different purposes. It strengthened the established rapport further. It helped in

Table 3

Summary of Data Collection Procedure

	Interview Format	*Tool*	*Type of Informant*	*Purpose*	*Background of Investigator Collecting Data*
1.	Structured Form	Village Profile *Anganwadi* Data	Key informants: Village level worker (Gramsevak), Talathi (local official maintaining land records), local leader (Sarpanch), other elderly persons in the village, *Anganwadi* workers.	To collect data regarding the village: demographics infrastructure and other services available. To collect data regarding the nutritional status of children.	Rural youth, the students of WOTR, trained for the purpose.
2.	Semi-Structured Guideline	FGD	Women: Mothers and caregivers of children 0–5 years	To collect data on community perceptions and practices.	A team of anthropologists and social workers by training.
		IDI and IC	Elderly women and men from community, Field officials of WOTR and SAMPADA TRUST Key Informants: In addition to the group of informants as mentioned above, here the list includes Village Watershed Committee (VWC) members, health promoters, local leader, Police Patil, other people.	To understand the change from the perspective of community and the field officials of NGO, WOTR and SAMPADA TRUST.	
		OB	Village and people in their settings were observed	To observe the practices in relation to community perspective.	

Note: *Anganwadi* is well known. It is a centre through which the government Integrated Child Development Services (ICDS) are implemented. The services cater to the pre-school children, age 0–6 years, pregnant and lactating women.

cross-checking and actually observing the life of the people and understanding the facts.

Informed Consent

Consent at the Level of Village

Introductory village level meetings were conducted in all the three villages before beginning with the data collection. The village meetings were fixed up about a week in advance by consulting people. The officials of WOTR and the social worker associated with the study helped in coordinating meetings with the people. In the meetings the study was introduced, the purpose of study, selection of the village, aspects to be covered in data collection, accommodation arrangements and others were discussed with the people and their consent was obtained to carry out the work in the villages. The people were keen about the benefits of the study. It was explained to them that there were no direct and immediate benefits of the study as such. The study would contribute to the society in terms of knowledge generation. Knowledge generation contributes to change in the perspectives at different levels: researchers, policy makers, development professionals, local leaders and community contributing to village development. The local leader, sarpanch, the village level worker, gramsevak, VWC members, school teachers, *Anganwadi* workers, elderly men and women of the village were present for the meetings. The people had been very supportive to the study all over and made arrangements for the researchers to stay during the period of data collection. It was explained to the people that it would consume some of their time for the purpose. The time would also be fixed up as per people's convenience and consent. Consent was also sought for photography of village scenes, people in action and recording of interviews and discussions. Some of the issues related to children like data on child mortality could be sensitive, but were important to document from the point of view of the study.

Consent at the Level of Informants

An informed consent form was developed to obtain consent of the informants before the actual interview and/or focus group discussion. The consent form initially developed in English was translated into local language, Marathi. The following components were included in the informed consent form:

1. Introduction of the researcher, organisation and the topic of study
2. Study procedure
3. Benefits of the study
4. Risks
5. Confidentiality of information obtained from interview
6. Voluntary participation
7. Required time of the informants
8. Use of audio-visual methods

There were no specific risks associated with interviewing and most took place in an informal manner with a freedom for openness. The conversations took place while the informants continued doing their household chores. FGDs were planned in advance and were conducted in household premises. During planning, the importance of the selected participants for FGD and absence of other members was explained to the people and the family members in a friendly manner and accordingly support was rendered. The FGD was conducted by two persons. One facilitated the discussion and the other recorded observations. The interviews and FGDs were tape recorded.

Before starting with the interview or discussion the study was briefed out to the informant/s, the consent form was read out and oral consent was obtained. It was decided not to obtain the signatures of the informants on the consent form as it would obstruct the friendly relationship, a kind of openness that exists between the researcher and the informants. On the contrary, the interviewer signed on the consent form as confirmation of reading it out. A card having name and full contact information of the researchers and the organisation was given to all the informants who participated in the interviews and focus group discussions. For informal conversation, no individual consent was obtained. It was a general informal talk not followed as per the interview guidelines and not always related to the objectives. There was no specific guideline for informal conversations. But the data elucidated helped in analysis and interpretation.

Data Analysis and Interpretation

The tape recorded interviews and discussions were transcribed. The data were not translated until the coding was done so as to catch the exact meaning. Translation was done in English in the first step of analysis.

The qualitative data analysis is done manually. Initially, coding was done to classify the data under different broad categories. The data were then organised and reorganised in several steps, to view the findings in clear and refined form. The data analysed have been interpreted in the context of watershed development inputs. As the study is comparative, the analysis illustrating the similarities and differences was cumbersome. As the refined data were put in a matrix form, a comparative picture could be obtained. Interpretation followed the analysis and the whole was seen in the context of watershed development inputs and the changes occurring in different communities. The similarities and differences were understood at two levels, one at the level of three comparative groups: villages 1, 2 and 3 and second was at the level of communities. The two main communities found consistently in three villages are the caste community Maratha consistently with predominant population and the tribal communities, the Thakar, the Bhil and the Lamani in Mhaswandi, Kangar and Kalas Pimpri respectively. The changes with similarities and differences had to be analysed across the communities too. The interpretation is put in a community perspective. The analysis of change among the two different communities was important to understand how different sections of the population were benefiting from the development programmes.

The data on the nutritional status of children obtained from *Anganwadi* is analysed using Epi Info™ Version 3.5.1. Epi Info™ is a public domain software package designed for the global community of public health practitioners and researchers. It provides for easy form and database construction, data entry and analysis with epidemiological statistics, maps and graphs (http://www.cdc.gov/epiinfo/downloads.htm).

Definitions of Terms as Used in Research

The terms and concepts used in the research work have been defined operationally as follows:

Watershed Development Inputs

It includes inputs of two types, the watershed development and participatory growth monitoring.

Watershed Development

This is defined in the text. Watershed development involves trapping and harvesting rain water with ridge to valley bases in order to make water available year round to all the people living in a particular area. It encompasses comprehensive, integrated and participatory as core components based on the approach of IGWDP.

Participatory Growth Monitoring

It involves inputs of monthly monitoring of weight and height of children 0–5 years, feedback to the community and involving the mothers of the children in the assessment of nutritional status of their children, holding of balanced diet demonstrations and healthy baby competitions to create an awareness of the nutritional food and status of children 0–5 years. The activity is facilitated by an NGO. This does not include the growth monitoring undertaken by Integrated Child Development Services, the *Anganwadi* centres. The village Kalas Pimpri selected as No Input has ICDS input but no other growth monitoring inputs as defined.

No Input

It is important to understand the category of 'No-Input'. No Input is in relation to watershed development and growth monitoring. If watershed development based on IGWDP approach with integrated, comprehensive and participatory component exists it is referred to as watershed development input village. If not in this sense, it is considered as no watershed development input. If growth monitoring input in the sense as defined above is absent, it is referred to as no growth monitoring input.

Culture

Culture is that complex whole which includes knowledge, belief, art, morals, law, custom and any other capabilities and habits acquired by man as a member of society (Tylor, 1924). It depicts a world view of the informant. Culture differs from one community to the other.

Change

It refers to variation.

Culture Change

A change in perceptions, that is, systems of knowledge and beliefs is referred to as cultural change.

Social Change

It refers to any event or action that affects the way of life of a community. It is the same as cultural change but is used as a wider concept.

Perception

Perception is simply defined here as one's own understanding or the information one has about a topic.

Nutrition

Here it refers primarily to the process and pattern of food and milk consumption, feeding practices.

Child Nutrition

It refers to the nutrition of pre-school children, age 0 to 5 years. The terms 'growth' and 'health' are associated with the term 'child nutrition' and are used in relation to weight and height of children 0–5 years.

Age Group 0–5 Years

This age group refers to age from birth to age five years complete.

Violence

The term refers to anger, physical assault like hitting and/or pushing and threats of the same.

Abuse

Abuse here refers to use of bad words or language towards a person.

Commercialisation

The term refers to the act of making parts of land, availability of water for more profitable and productive purposes. It refers to exploitation of resources of land and water to that extent.

Small Farmers

Based on field experience, the term refers to farmers owning land less than or equal to four acres.

Medium Farmers

Refers to farmers owning land more than four and upto eight acres.

Large Farmers

Refers to farmers owning land greater than eight acres.

Figures Highlighting Study Site

The study site is highlighted on maps.

Figure 2

Map of India with State Maharashtra Highlighted

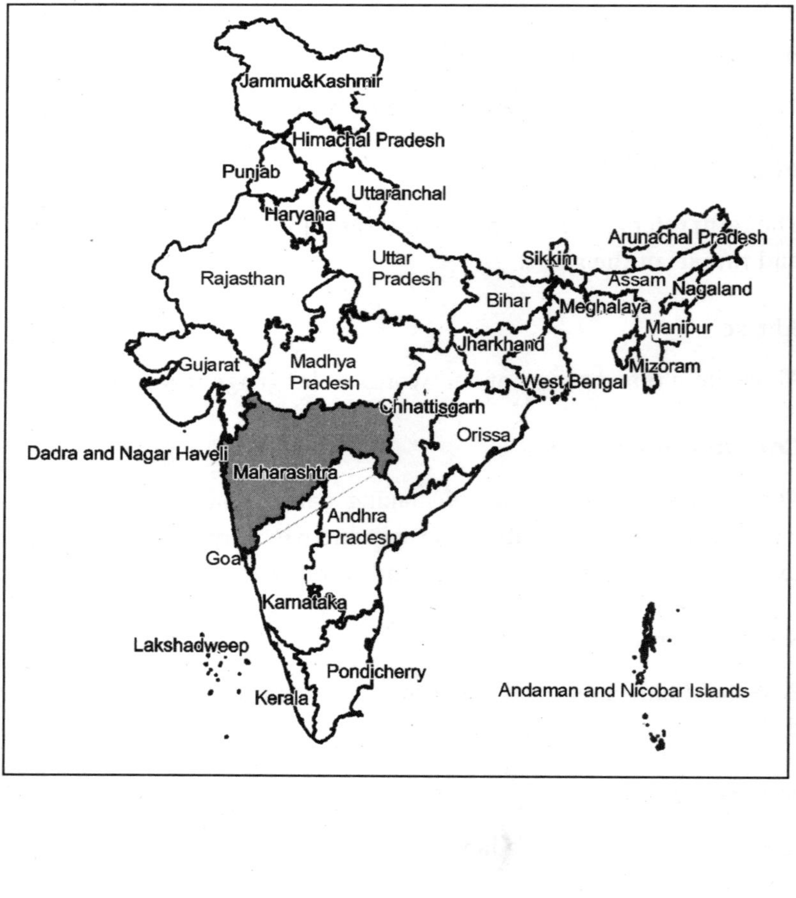

Figure 3

Map of Maharashtra with District Ahmednagar Highlighted

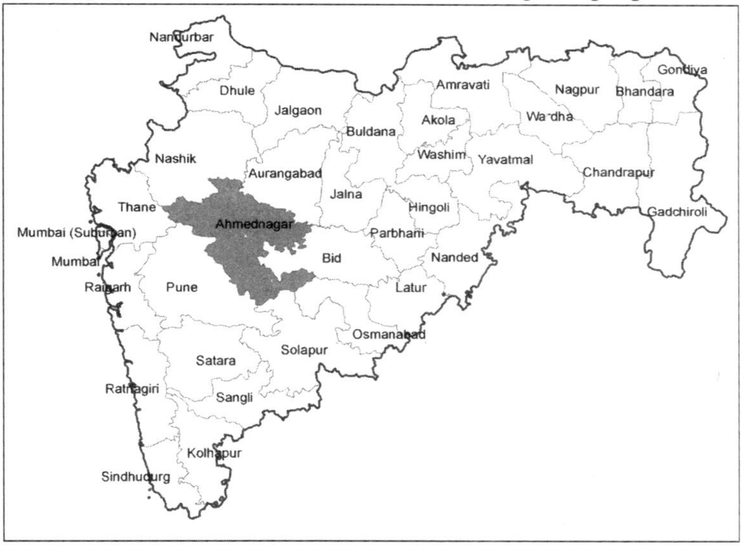

Figure 4

Map of District Ahmednagar with Study Talukas and Villages Highlighted

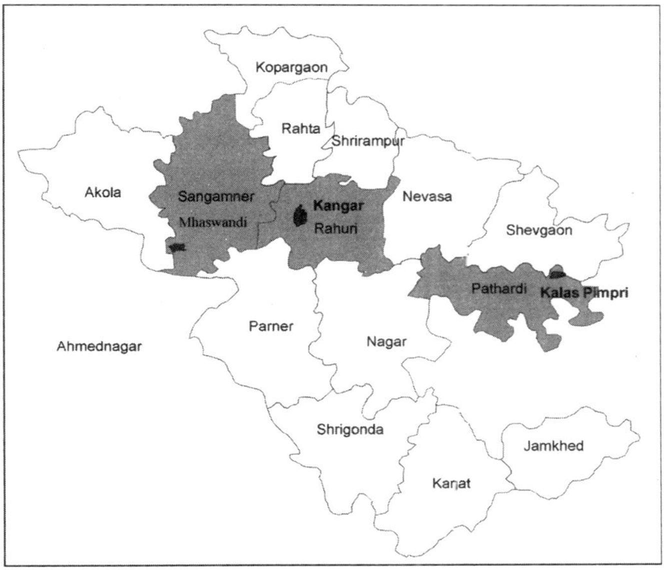

3

Studied Villages: A Profile

The studied villages belong to a drought-prone district, Ahmednagar in Maharashtra. Agriculture is the primary means of livelihood in all the three villages. Pearl millet/Bajra (Bajri is the local word used in Maharashtra) is the staple crop. Onion *(Kanda)*, tomato *(tambate),* potato *(batata),* groundnuts *(Bhuimug)* are the cash crops. Onion offers huge profits in watershed areas. In Kalas Pimpri, cotton *(sarki)* is the main cash crop. Onion, tomato, potato, groundnuts are grown in meagre amounts in Kalas Pimpri as there is not enough water for the cultivation of these crops. Kangar is a large, spread out village than others. The villages are similar in terms of ethnic composition, mode of living and environment characteristics.

Mhaswandi (WSD+GM)

Mhaswandi represents the watershed development and growth monitoring inputs group, henceforth referred as WSD+GM group. It is a village located in Sangamner taluka of Ahmednagar district. The village is located in a hilly area and is surrounded by mountains. The presence of streams and canals, onion and bajra fields all over make

Plate 2

Watershed Village

the village look green and beautiful. Watershed Development (WSD) has been a great success in the village. Once possessing dry land, it has been converted into irrigated land after the watershed development. The village has been free from tanker service *(tanker mukta)* for drinking water. Water shortage is reduced to a large extent. The shortage is severe primarily for two months in one of the hamlets which is on hilltop. There is plenty of water available in the village than before for domestic consumption, agriculture and livestock. There is greenery all over the village. Floriculture is well-developed.

The villagers are particular about sanitation and hygiene. The practice of open defecation is discontinued by the people and a penalty is taken if found. The village is on panchayat record for being free from open defecation (*Hagandari mukta*). Latrines have been constructed by many households. It is a joint decision to make use of the latrines and a fine is levied for open defecation.

One does not find poultry much in the village. The villagers have decided not to keep poultry. The poultry is believed as spoiling the cleanliness of the village.

Kangar (WSD)

Kangar represents a village with watershed development input. There is no growth monitoring for children 0–5 years done by any of the NGOs in the village, henceforth referred as WSD group. It is a village located in Rahuri taluka of Ahmednagar district. It is a large village spread over a wide distance. The village has been largely benefited from watershed development in terms of the availability of water for a large part of the year. Water shortage has greatly reduced. There is no water shortage for drinking in summer, said the people. Bajra/Pearl millet is the staple crop. Onion is the main cash crop.

Plate 3

An Evening in Watershed Village

Tomato and groundnut are grown as the other cash crops. The village exhibited a striking difference from Mhaswandi. The invisibility of lush greenery and a wide variety of crops contrasted with that of Mhaswandi. Floriculture is not practised. Here dairy forms an important source of income. Rahuri Sugar Factory that is about seven to eight km from the village is a good source of employment for many.

At the entrance of the village, along the roadside one finds many small shops such as cycle and puncture repair shop, electrical shop, small hotels, grocery shop and mutton shop.

The village has a library with a good collection of books. There are story books for small children in the library. The library is open daily from 4:00 to 7:00 pm.

A mobile clinic of Pravara Medical Trust visits the village once in a fortnight. People get medicines at low cost. They charge two rupees per person for medication.

Kalas Pimpri (No Input)

Kalas Pimpri represents a village with No Inputs of watershed development and NGO initiated participatory growth monitoring, henceforth referred as 'No Input' group. The village lies in Pathardi *taluka* of Ahmednagar district. The whole of Pathardi taluka is dry. During the period of data collection, the village fields looked dry and brown that contrasted with Mhaswandi. There is cotton all over the village. It is the main cash crop in the village offering huge profits. Cotton is grown in large amounts in the most of Pathardi. Pearl millet is the staple crop. In discussions people often mentioned that the dry land in the village is not so productive to grow a variety of crops.

There is heavy migration for employment outside the village. Sugar factories and coal mines in Karnataka and Rajasthan are the places where people migrate for labour work. Work in coal mines gives higher returns.

A government residential school *(niwasi yojna)* exists in the village to cater to the education needs of the children of migrating families during the period of migration. The school provides

Plate 4

Commercial Crop in the Dry Land

residence to the children in school until the parents are back from migration.

Village Profiles

The profile of the three villages, representing three comparative groups, viz., Mhaswandi (WSD+GM), Kangar (WSD) and Kalas Pimpri (No Input) is put down in a matrix so as to have a comparative understanding at a glance. The interpretation of the table is given at the end of the matrix.

Table 4

Village Profiles in Comparison

	Features	*Villages*		
		Mhaswandi	*Kangar*	*Kalas Pimpri*
I.	***Physical***			
1.	Number of hamlets	6	12	5
	Distance in km from			
2.	District	100	35	75
3.	Taluka	50	12	25
II.	***Population (persons)***			
1.	**Total**	**1,165**	**4,389**	**1,187**
1.1.	Number of women	581	2,248	584
1.2.	Number of men	584	2,141	603
1.3.	Sex ratio: total	995	1,050	968
1.4.	Total children: 0–5 years	122	421	147
1.5.	Number of girls	60	215	79
1.6.	Number of boys	62	206	68
1.7.	Sex ratio: 0–5 years	968	1,044	1,162
2.	***Ethnic groups***			
2.1.	Maratha	762	2,900	633
2.2.	Thakar	339	0	0
2.3.	Bhil	19	508	0

Cont'd...

Cont'd...

2.4.	Lamani	0	0	291
2.5.	Mahar	16	787	171
2.6.	Others	29	194	92
III.	***Communication***			
1.	Transport services available to taluka from a place (distance).	4 km	0 km	2 km
2.	Frequency of transport to taluka place in a day.	once	more	more
3.	Post office	No	Yes	No
4.	Telephone facility (public, private, mobile)	Yes	Yes	Yes
IV.	***Services: distance where available***			
1.	Health	5 km	0 km	0 km
2.	Veterinary	10 km	4 km	5 km
3.	*Anganwadi* (ICDS)	1	4	1
4.	Education			
4.1.	Primary	0 km	0 km	0 km
4.2.	High School	5 km	0 km	2 km
4.3.	Degree college	18 km	12 km	21 km
V.	***Shops***			
1.	Medical shop	No	Yes	No
2.	Ration shop (Public Distribution System)	Yes	Yes	Yes
3.	Grocery store	Yes	Yes	Yes
4.	Liquor shop	No. Banned by women	Yes	Yes
VI.	***Market: distance where available***	10 km	0 km	5 km
VII.	***Gram Panchayat***			
	Type	Single	Single	Single
	Total members	7	11	7
	Women members	2	4	2

Cont'd...

Cont'd...

VIII	***Dairy/milk cooperative societies***	One society with milk collection approx. 400 to 500 litres per day (lpd).	Three societies with milk collection approx. 15–17,000 lpd.	No milk cooperative. Private milk collection but not a dairy, with collection of approx. 115 lpd.
IX	***Watershed development***			
1.	Area treated	1,145 hectares	3,000 hectares	NA
X	***Growth monitoring***	Mhaswandi	Kangar	Kalas Pimpri
	Year	June 2006	Not initiated	NA
XI	***Women institutions***			
1.	Self-Help groups	23	12	2
2.	Women's committee (Samyukta Mahila Samiti)	Yes. One.	No	No

Note: The category 'others' in the population includes communities whose occupation in the traditional Hindu hierarchy is given in bracket: *Teli* (Oil Pressers), *Beldar and Wadar* (Stone workers), *Sutar* (Carpenter), *Nhavi* (Barber), *Brahmin* (Priest), Muslims, Pardhi and Mang.

Discussion

Physical Features

Among all, Kangar is a large village. The population and the number of hamlets are more than the other villages. Mhaswandi and Kalas Pimpri are comparatively smaller. In terms of population and the number of hamlets, they are of similar size.

Sex Ratio

The sex ratio of total population is highest in Kangar with Mhaswandi ranking second and lowest in Kalas Pimpri. The sex ratio of children 0–5 years is highest in Kalas Pimpri among all with Kangar ranking second and is lowest in Mhaswandi, in comparison.

Ethnic Composition and Analysis

In all the villages, there are two main endogamous groups, caste and the tribe. The caste Maratha outnumbers the others in all the villages. Mahar, another caste group ranks second in population in village Kangar. While the tribal communities rank second in population in two villages, Mhaswandi and Kalas Pimpri. In Kangar, they have considerable population but rank third in population following the Mahar. There are three different tribal groups, viz., the Thakar in Mhaswandi, the Bhil in Kangar and the Lamani in Kalas Pimpri. The ethnic composition is such as a dominant caste community like Maratha at the one end and the tribal community at the other end found in all three villages. Culturally, socially, economically they are at two different ends. With this ethnic composition at background, the data are collected from the Maratha, the tribal communities and the caste community Mahar. As the Mahar are found in more numbers in one comparative group and their population in Mhaswandi is too less, a three-tier comparative analysis of Mahar was not possible. The comparative analysis and discussion would focus on two communities, the caste community Maratha and the tribal groups as mentioned before.

Infrastructure: Communications, Market, Services Available

Mhaswandi is a village located in a hilly area. Its connectivity with the outside world is comparatively weak. For example, communication services available are less. The village is more at a distance from district and taluka places than other villages. Market is far. High school education is not available in the village. Degree education is distanced far and with poor communication, the accessibility is affected. Unlike other villages, health services, medical drug store, veterinary services are situated far away from the village. The infrastructure of Mhaswandi is comparatively poor. Kangar has a strong infrastructure. It is comparatively close to district and taluka places. It ranks better in terms of the availability and accessibility to communication, health, medical, education services. Market is right there in the village.

Livelihood

Agriculture is the primary means of livelihood in all the villages. Since watershed development, the average income of the villages has risen to a large extent. For example, income from cash crops, onion, tomatoes, floriculture in Mhaswandi, dairy farming and earnings from cash crop onion in Kangar have been related to the development of watersheds in the villages by the people themselves. Regarding dairy as a means of earning, Kangar fares well among all. It has large milk collection and there are three milk cooperative societies. While in Mhaswandi, till a year ago, the village had an annual income of 1.5 crores from onion production.

Women Institutions

Among all villages, Mhaswandi has more number of self-help groups (SHGs). The SHGs have federated into a larger women's committee, the SMS. SHG, SMS are people's institutions. Watershed development has initiated and strengthened the people's institutions and Mhaswandi and Kangar are examples in case. There is a good number of SHGs in Kangar too but less than that of Mhaswandi. In Mhaswandi, the initiation of development activity of participatory growth monitoring of children by Sampada Trust facilitated continuation of development inputs and NGO contacts which is not the case in Kangar as there is no growth monitoring in Kangar. Also as the main activity of Sampada Trust is women empowerment, women institutions assume great importance. These factors seem to be contributing to more number of SHGs and the formation of SMS in Mhaswandi as against Kangar. In Kalas Pimpri, there is no watershed development or growth monitoring by Sampada. People are interested in forming SHGs and they expressed the need for SHG inputs. The village is in the process of formation of self-help groups.

4

Watershed Development Inputs and Cultural Change

The present chapter discusses social changes as influenced by watershed development inputs in the input villages, viz., Mhaswandi and Kangar. Watershed development inputs include two major input components as discussed before:

1. Watershed Development
2. Participatory Growth Monitoring of Children 0–5 Years

The social changes associated with watershed development inputs are studied with regard to both the above components. The following are some of the aspects where changes are observed:

1. Agriculture
2. Livestock
3. Environment
4. Employment
5. Community
6. Women
7. Health

Watershed development in Mhaswandi and Kangar villages has resulted in agriculture, livestock and other developments and allied income opportunities.

The caste community Maratha community population is more than the tribal Thakar and Bhil communities in the watershed villages. The Maratha possess more land. The Thakar and the Bhil possess meagre and fragmented land, usually not exceeding four acres. Many are landless too. They reside on the hilltop and possess land on the hilly terrain. The Maratha predominantly reside in the plains and comparatively possess more of fertile and irrigated land.

Pearl millet is the staple crop. Crops like red gram dhal *(Tur)*, green gram dhal *(Mug)* are grown in considerable proportion. Wheat, *Jowar*, paddy are grown in small amounts. Wheat and garlic are grown by tribal families for subsistence. Onion, tomato, potato and groundnut are the cash crops. Onion is one of the most important commercial crops, leading to profit in crores. It is grown in three seasons, i.e., *Kharif*, late *Kharif* and *Rabbi*.

In the present chapter, the impacts of watershed development inputs on community life and its implications for child nutrition have been discussed. The impacts are broadly discussed in terms of benefits to the villagers and limitations of the programme.

Watershed Development and Social Change

Agriculture

Crop Assurance

The discussions with people and personal observations indicate that prior to watershed development the crops were dependent on rains only and hence there was no guarantee of the crop yield. After watershed development, a change has come about. There is surety about the crop yield due to irrigation. The tribal people in Mhaswandi and Kangar have been able to take new crops like tomato, onion, garlic, wheat and potato. For the tribal small farmers this led to an increase in crop variety serving their livelihood needs.

Increase in Cultivable Land

After watershed development, the area under cultivation has increased. Due to land levelling many farmers could cultivate more area of land as compared to earlier situation. Earlier only those farmers whose land was on the plains or whose farm soil quality was good would take the second crop.

New Varieties of Crop: Food and Fodder

Prior to watershed development due to lack of alternate sources of water, rainfed farming was the only option. The Maratha as well as Thakar farmers cultivated only *Kharif* crops like pearl millet, pulses *(kaddhanya),* groundnut, gram (*harbhara), shaalu* (probably a millet). After the watershed development, farmers have also started cultivating crops such as onion, tomato, potato, garlic, wheat, paddy and sugarcane.

In Mhaswandi, besides onion, potato is grown in large amounts. People have started growing *Rabbi* crops (winter crops) due to availability of water for a longer period of time. The Mhaswandi villagers were seen practising floriculture since the last five years. Flower cash crops such as marigold *(zendu),* aster, chrysanthemum (*shevanti)* and *bijli,* another flower cash crops (English name not found) are grown. More number of Maratha as well as few Thakar families too practice floriculture.

Plate 5

Floriculture in Watershed Villages

Onion, tomato, potato, groundnuts are the cash crops that offer huge profits. There is huge production of onion. In the past one year, Mhaswandi earned 1.5 crores from onion production. Onion production is measured in truckloads and tonnes in Mhaswandi. The village makes use of sprinkler system and drip irrigation for onion crops.

Due to more availability of water for agriculture, farmers can also take fodder crops and grasses like corn and *ghas* for animals. This fodder is used as stall feed for the cross-bred animals. The fodder crops are mostly grown by the Maratha in the watershed villages. It was observed that fodder crops were not grown by the tribal communities in watershed villages. Not much of fodder crops were seen growing in No Input village, Kalas Pimpri. As very few agriculturist families in Kalas Pimpri have cross-bred cows they do not feel the need of cultivating fodder crop as stall feed.

Increase in Crop Yield

Increase in cultivable land resulted in increase in crop yield. After watershed development, the crop production has doubled. The situation is similarly found among all communities. When compared between the caste and the tribe communities, the total production is comparatively less among the tribal communities. One of the reasons for which is their land conditions: less ownership, fragmented land and land on hilly terrain. In Kalas Pimpri, pearl millet and cotton are grown in large amounts and due to dry land conditions not many crops are taken. Onion is grown in very less amount. People in Kalas Pimpri mentioned that they cannot grow onion because of inadequate availability of water. Tomatoes were not found. The difference in cropping patterns and the yield is obvious and prominent as one moves around in the three villages

Surplus

Prior to watershed development, there was no commercial yield, said the people. Whatever was grown was sufficient for domestic consumption with a little surplus for distribution. However, after

Plate 6

Watershed Village: Tomato and Onion Production in Tribal and Caste Families Respectively

watershed development, the crop yield increased to a large extent. Examples: In Mhaswandi and Kangar, large amount of onion is sold out in the market. Nearly 200–300 truckloads of onion have been sold in one year with resulting income in crores. Pearl millet is usually left for domestic consumption with all cash crops sold out. The markets of Narayangaon, Aale phata, Otur, town places are close to Mhaswandi and Rahuri tehsil market is closer to Kangar, all distanced at the range of 20–35 km. The people said that prior to watershed development, they went outside villages for purchase of onions and potatoes. It was observed that the Maratha concentrate on onion production and sale and the tribal small farmers on tomato production. Tomato production is measured in local measuring unit, *carat*. One of the tribal informants, a small farmer with about four acres of land in the hilly area of Mhaswandi village had sold about 30 *carat* of tomatoes. A pond that was constructed nearby his house was a source of irrigation for him. One *carat* of tomatoes equals to 20 kg which means he sold out 600 kg of tomatoes.

Livestock

The local/*desi* cows are replaced by cross-breds in the villages. More number of Maratha possess cross-bred cows. The cross-breds require stall feeding. Stall feeds are available due to fodder crop production. This has led to rise in milk production and dairy farming has increased.

More number of tribal Thakar from Mhaswandi and the Bhil from Kangar possess local cows, goats and buffaloes as their livestock while the Lamani from Kalas Pimpri keep local cows and goats.

Availability of Fodder and Water

Since watershed development, the scarcity of fodder and water for the livestock has been reduced. As the culture of raising the fodder crop in the farm has crept in, a sufficient quantity is available for not only the animals but there is enough surplus for sale. This was not the condition before. People had more number of local cows which grazed outside. There is increased availability of water ensured for

Plate 7

Dairy in Watershed Village

the domestic animals, the cattle and the goats all round the year in Mhaswandi and Kangar.

Dairy

Milk production has increased. Earlier there was milk production from local cows and milk was utilised mainly for domestic consumption. There was not much of surplus milk for sale. With watershed development, rise in fodder production, purchase of cross-bred cows, milk production has shot up. In Mhaswandi, the villagers have purchased a *vankular* machine, a machine in which upto 500 litres of milk can be stored for a day in case the collecting van does not come to the village. In Mhaswandi, daily milk collection is upto 400–500 litres. In Kangar, it is about 15,000 litres. A comparison among the Maratha from three villages shows very few

Maratha farmers from Kalas Pimpri village have cross-bred cows. There is no dairy in Kalas Pimpri. One person collects milk from the village and sends it outside.

Environment

Increase in the Number of Wells

Prior to watershed development, all the rain water would just flow out, said the people in conversations. After land treatment, the water flow is blocked and it then percolates in the soil. Consequently watershed development leads to increase in ground water table, said the people. The people in the villages have dug more wells in the farms. The water is used for irrigation and for drinking in case there is a shortage in the months of summer.

Increase in Green Cover

Prior to watershed development, the villages looked brown and dry. In summer months, there was complete dryness in the area. During the watershed development period, for example in Mhaswandi, five lakh trees were planted in the hills and waste land (trees such as *Subabhul,* eucalyptus were planted). Tree plantation resulted in increase in the green cover of the village, said a woman informant.

Increase in Rains

Due to plantation of many tress and resulting greenery, the amount of rainfall has increased as experienced by the people from the villages.

Fauna

The number of small and few large size birds has increased. During our stay in the field we too noticed variety of birds especially in village Mhaswandi. The people mentioned that they actually have visualised the change. Due to increase in greenery, the number of peacocks has increased. Animals like fox, wolf, deer and wild pigs are seen in the villages as mentioned by the people.

Drinking Water

The scarcity of water in the months of summer was severe prior to the watershed development. There is reduction in water scarcity. Out of the four summer months, water scarcity has reduced from four to two months in summer only in the hilly parts of the village. In Mhaswandi, all hamlets except the hilltop hamlet have water throughout the year. There is no scarcity even in summer. In our field visits to the hilly tribal hamlets, we found that the tribal and few Maratha living on the hilltop have to face water scarcity for two months, almost from April end upto the commencement of rains in the month of June, which was much severe before. As against Mhaswandi, in Kangar there is no water scarcity in summer and water is available for drinking throughout the year. Kangar is comparatively situated on a plain area. Water is available for a longer period of time, which was not the situation before. Women had to travel long distances for washing clothes. They spent two to three hours in fetching drinking water and washing clothes. In summer, most of the day was spent in finding and fetching water, said the people. During the period of water shortage people had to cut short on water for bathing. As piped water connections are now available inside houses the situation has changed. More water is available for bathing. The tribal people too get drinking water close to their houses.

Prior to watershed development, there was heavy dependency on rain water. Water scarcity was a major problem in summer. Wells dried up in summer months from February to June. People travelled two to three km in search of wells and water. If water was not available in the wells, tanker services were availed. The tanker water was filled in the wells in the hamlets. It was a great struggle to obtain the tanker water, said the people. Villagers went to the Panchayat Samiti office, stayed there for long hours, fought and brought the tanker to the village. Since watershed development, the problem of requesting for and struggling to obtain tanker water has minimised. Tanker service is no more a requirement for drinking water as earlier during summer in the two villages. The people mentioned that about eighty per cent of families obtain drinking water from their own wells. The people from central part of the village, the *gaothan* get piped water.

Economy

Benefit to Nearby Villages

Watershed development has benefited all the people from Mhaswandi and Kangar watershed villages. The villages in proximity of 4–8 km have also benefited in terms of more availability of water for drinking and agricultural purposes. Villages close to Kangar like Ganegaon, Chinchavare have got the benefits and villages close to Mhaswandi like Ambidumala, Pimpri and Jachakwadi, which are situated on plains have benefited. Mhaswandi is a hill village and the villages nearby Mhaswandi are on the plains. The geographical conditions have been favourable to get the benefits from watershed development. The people from nearby villages have visited the watershed villages with great interest. During a field visit to the surrounding villages, they gave a feeling of more agricultural prosperity than even Mhaswandi, a watershed village. The village Ambidumala for example has huge orchards of pomegranate *(Dalimb)*. More number of permanent houses made up of cement concrete were found in Jachakwadi. Onion production seemed to be huge in Jachakwadi.

Availability of More Work

The Maratha opined that under the watershed development programme, initiation of soil and water treatment activities generated labour opportunities in the village. As the villages around watershed villages have also received the benefits of watershed development, more of agricultural labour work is available than before. The tribal small farmers who earned their supplementary income from labour work by migrating outside the village find agricultural labour work in close proximity to their resident villages. This has reduced migration of tribal people during the period when they have less or no work in their farms. Prior to watershed development, during the period from January to June, in watershed villages, the tribals as well as the Maratha migrated out in search of employment. They went to far off places to work as agricultural labourers, or as labourers in sugar factories and/or brick kilns and returned just before the rains. As the benefits of watershed development were being realised, farmers started taking more than one crop in a year. The scope for labour work such as weeding, sowing, and harvest increased. Agricultural work is available for at least eight months in a year. The number of people moving out in search of work for six months has declined since the availability of agricultural work. The tribal people mentioned about the availability of more of agricultural labour work and increase in work in their own farms after watershed development. Migration of villagers (whose land became irrigated after watershed development) has reduced when compared with the earlier situation. People who possessed land migrated out to cities like Mumbai in search of work but watershed development generated interest in their own farms.

Increase in Wage Amount

The landless poor are paid for their labour under the watershed development work. In the watershed villages they obtained a daily wage rate of rupees hundred for working as labourers in watershed development projects. The daily wage rates have increased since then and the same practice is existing thereafter for other labour work they get in and around their villages.

Improved Standard of Living

The people believe that due to the increase in farm yield, the economic condition of the villagers has improved. This in turn has resulted in improvement in living standard. That they can now purchase food items they want for their children. One of the Maratha caste informant said that the tribal people can also take care of children in a better way. This has led to happiness in life *(Te ek sukh tyana zale)*.

Due to availability of money, now the people do not purchase items on credit (*udhari*) from the shop. Purchase of items by cash is now possible due to prosperity.

A notable change in the standard of living among the tribal people is found in personal hygiene. The discussions revealed that as the streams have plenty of water; bathing has become a regular feature for the children in particular.

Village Harmony

Mhaswandi is a good example exhibiting village harmony. Factions existed in the villages. Often there were quarrels. The quarrels were more during the elections. There were three liquor shops in the village. The business of money lenders (*savkar dhanda)* was in full swing. After watershed development, a ban on liquor shops was imposed by village women. The SHG women from the village unanimously fought towards ban on liquor and money lending. The money lenders are not allowed to practice any more. The self-help groups of women cater to emergency fund raising.

In Mhaswandi and Kangar, inter and intra community relationships are fostered with the initiation of watershed development. During implementation the village united and contributed voluntary labour following the three conditions of watershed development that are: ban on grazing, ban on tree-cutting and contributing voluntary labour. The interactions and relationships between the tribal and the caste communities have improved.

Watershed development initiated building up of local institutions. The village union is strengthened with formation of various committees: the Self-Help Groups, Village Watershed Committee, and the milk cooperatives. The tribal people are also members of VWC. The problems in the villages are solved unanimously. The village meetings (gram sabha) are conducted regularly as and when the need arises. In Mhaswandi, there is unanimous election (*bin virodh nivadnuk*), may that be for panchayat leadership or committee presidentship or any other sort of election for the past few years.

The local institutions are strong in Mhaswandi than in Kangar, the reason discussed as post watershed NGO contacts and continuation of development activities. The number of self-help groups and the federated body is an example of the same (reference: village profile). People talked of self-reliance in their life and satisfaction after the watershed development.

Livelihood Issues of the Tribal People

For the tribal people, watershed development has brought about a change in livelihood patterns with positive and negative effects.

The increase in agricultural work in the villages has offered employment opportunity to the landless tribal who have started working on the farms of the large farmers on share basis *(ardhali)*. The yield is divided equally between the farmers and the agricultural labourers. However, the ban on which the watershed development is implemented have hit the tribal people worst. A part of the tribal people's livelihood need was based on collection of forest wood. Watershed rules like ban on free grazing, ban on tree-cutting have restricted their access to resources. The tribal people kept goats for domestic milk consumption, meat and sale. Due to the ban on free grazing, they have started facing difficulty in rearing goats. They have to cover the mouths of the goats whenever they move in the village. The goats need constant attention while they work on the farms so as to prevent them from free grazing increasing the drudgery of tribal women. Many have sold out their goats and have lost interest in goat rearing as its management has become difficult. Similar is the situation with regard to local cows. The tribal people are giving up rearing local cows. Unlike the cross-bred cows, the local cows need movement and cannot be stall fed. Due to ban on free grazing and the penalty for breaking the rules, the tribal people have started selling out their local cows too affecting their livelihood.

The dominant caste exercises control on natural resources like free grazing, use of water resources and domestication of animals. In Mhaswandi, with watershed development exhibiting economic growth people have become more conscious of social status and prestige of their village due to prosperity and visitors from outside places visiting the village for observing the impact of watershed development. People have taken great interest in environmental sanitation. Latrines have been constructed and there is a ban on open defecation. A penalty is levied on open defecation. They want the village to be known as free from open defecation *(Hagandari mukta)* and be addressed as ideal village *(adarsha gaon)*. The Maratha have also made a rule that there should be no poultry in the village as poultry creates dirt *(kombdya khup ghan kartat)*. It may be good to strive for reputation of the village from Maratha community perspective. The problem is the tribal community, a part of the village is not involved in the process of decision-making for village development. The concern for social equity has been lost in the process. Poultry is very much a part of tribal culture and livelihood. Poultry is required for rituals, meeting food needs and is a source of income for the tribal people. Another example is that a pond on the hilltop had been a good source of irrigation due to adequate availability of water for the tribal small farmers. They could make some profit out of their crop yield. A newly introduced fishery project in the pond by an outsider has restricted the use of water affecting the tribal livelihood. A tribal small farmer who earned some profit by selling 600 kg of tomato in the year may not have the opportunity the next year due to the fishery project. The village institutions, a part of watershed development are not sensitive to the issues of tribal culture and livelihood. It makes one desperate to find that social equity is remote to watershed development for it lacks the perspectives of the poor. Women institutions, of which the tribal women are also a part, raise no voices express regarding no concerns their livelihood issues.

Watershed development has a lot more to achieve in terms of social development. In this case, sensitising the communities towards their own issues and building capacities to resolve the same. It is not that watershed development is not benefiting the tribal people, but some social issues stand as obstacles in the development of the tribal people which need to be resolved unanimously by the tribe and the caste communities themselves. The development staff should be sensitive to these issues. The NGOs as facilitators can play a great role in creating sensitisation towards the issues of the poor in the process of implementation of watershed development. Watershed development has the potential to foster inter- and intra-community relationships as observed but the harmony needs to be strengthened further so that the problems of one community become the problems of whole village only after which they will resolve issues in union.

Women

Earlier the women went out of the village in search of employment. After watershed development, they are occupied in their own farms for a large part of the time. According to the elderly women and men, awareness of the women has increased with the formation of Self-Help Groups (SHGs) and *Samyukta Mahila Samiti* (SMS) which is a federation of SHGs as found in Mhaswandi. The men opined, "Earlier women were scared to even stand in front of men, let alone the issue of talking with them. But now after the WSD and SMS work they have become bold. Now they put forth their opinion freely even on radio or television. They have got the self confidence" *(Tyanchi vicharsarni badalali. Sankuchit pravutti badlun je kahi tyanna purushan samor bolana tar soda ubha rahanyachi suddha bhiti vatat hoti. Atta konihi purush yevo, baherchya rajyatil purush yevo, tya dhadhadhad boltat. Kivha mug kuthlyahi thikani akashwani aso doordarshan aso kuthlyahi thikani jahirpane apli mat vyakta kartat. Tyanchi bhiti je kahi aahe vyaspith tayar zala aahe tyatun tyanna sphurti milali. atmavishwas tayar zala)*. In one of the competitions of food preparation held by outsider group, the tribal women were the winners.

There are 23 women SHGs in Mhaswandi, out of which five have more number of tribal women. In Kangar, there are twelve women SHGs, with Maratha and few tribal women as the members of SHGs. According to the women SHG members from Mhaswandi, through the formation of SHGs, the women have their own little savings. The interaction between women has increased and they share each other's problems. The increased interactions amongst them and with outside exposure, a change is observed among the women. The women from Mhaswandi were found to be articulate and confident in voicing of their opinion as against the women from Kangar and Kalas Pimpri. During the focus group discussion with women from Mhaswandi, among both Maratha and the tribal communities, observation indicates Mhaswandi women are more open, more expressive in terms of putting across their thoughts regarding the topics discussed with them. They put forth their views and opinions confidently. They were aware of the changes that occurred in the village after the watershed development. They have successfully brought about a ban on liquor by struggling with village men. There were three liquor shops in Mhaswandi a few years back which were closed down by the women. The women also fought and obtained the rights of a woman who after her husband's death was denied her share of rights in household and agricultural property by in-laws. It was a great struggle she had to face against abuse and violence from in-laws because of husband's death and other family matters. She is now a health promoter in the village, trained by Sampada Trust. A change in the Mhaswandi village women was taken note by the media. Radio interviews, newspaper publicity, award grants have strengthened their capacities and confidence further. The women from Mhaswandi have great exposure to the outside public. There are visitors from schools, colleges, organisations, foreign students coming to view the village, to observe the change after watershed development.

The SMS promotes various entrepreneurial activities, viz., small scale industry wherein *masala* grinding machine and vermicelli (*shewai*) machine have been installed managed by a landless woman of the village, establishment of gas cylinder unit, cooking of supplementary food for the *Anganwadi* children, life insurance of all the women of the village, distribution of ration during the Diwali festival and distribution of loans to the needy women.

The self-help groups, the exposure, the entrepreneurship, the confidence, the courage to fight for justice among women of Mhaswandi, the WSD+GM study group is remarkable. The other villages, viz., Kangar and Kalas Pimpri, are yet to develop the way Mhaswandi shows the strength.

Observations and discussions during the focus groups with women from Kangar, a village with watershed development input alone shows a different picture. The Maratha women of Kangar are not much aware of the watershed development work carried out in the village and the resulting changes. They were not aware of the existence of a Village Watershed Committee (VWC) and two women as its representatives. The exposure of the women from Kangar was found to be comparatively less. The Maratha women said that they rarely went out of their hamlet to the central part of the village, called as *gaothan* which is about 4–5 km away from their hamlet. They questioned the researcher as to how would they know what happened in the village when they did not go out and worked only in the house (*Gharatlach karaycha. Kuthe jaycha nahi na aan yaycha nahi. Kasa mahiti hoil?)*. The male members from their family, especially the father-in-law and the husband attended the meetings outside and happened to go out. But they never discussed issues outside home matters and how would the women ask what subjects were discussed outside?, said a focus group informant (*Tyanna baherche vishay meeting vagere mahit hotat. Tar te gharat kahich sangati nahit. Baher kai vishay zala aaj apan kase vicharayche?)*.

The reasons for notable women development in Mhaswandi as against Kangar are partly because of continuation of post-watershed development inputs and NGO contacts. Kangar is a large village with dispersed settlements. The distance between the hamlets as well as between the houses is large. This poses a difficulty for the village women from different hamlets to come together or to even to move out of their houses.

Kalas Pimpri, a No Input village, exhibits a different picture altogether. There are two self-help groups in the village and their functioning is yet to become a routine. The meetings do not have a regular pattern as is found in other two villages. Here there is a difference in women from Maratha and the Lamani communities. Observations during focus group discussions with women indicate that the Maratha women spoke less, shied from talking to the researchers, spoke in low voices, took lots of pauses during discussion and required constant encouragement to speak out. By contrast, the tribal Lamani women were articulate and confidently expressed their opinions. They spoke in high voice and asked many questions from the researcher regarding their outlook on the topics discussed and solicited information on child nutrition and health. The adolescent girls, young women and elderly women exhibited a comparatively higher level of confidence. The Lamani migrate out to places like Karnataka and Rajasthan for labour work in coal mines for 6–8 months, where they earn in good amounts. Though a nomadic tribe, they have settled in Kalas Pimpri village. They have purchased land from the Maratha. Although their land is fragmented and rainfed, they make a lot of efforts growing cotton in large amounts. All of the cotton grown is sold out. The girls are encouraged for continuing their education. Young women are educated upto secondary level. The education of women and their exposure to the outside world here contributes to their world view.

Health

Improvement in Health Status

The Maratha of Mhaswandi perceive an improvement in the nutrition and health status of children with more availability of food and milk. However, the information came out on probing and not in open discussion. This may be because nutritional status is not perceived as an outcome of watershed development by the people. The Maratha from Kangar said that they did not know whether there is a change in the nutritional status of children after watershed development.

Change in Frequency of Meals

The people mentioned that watershed development has led to increase in the number of meals. Earlier, the villagers were not so keen on having early morning breakfast. Having full meal at mid-day, and early evening was a norm which has changed to three meals, breakfast, lunch and dinner.

Limitations of Watershed Development

In Mhaswandi and Kangar, the benefits of watershed development are unevenly distributed. More the land ownership, more the benefits. The large farmers have benefited the most. Most of the large farmers belong to Hindu caste community Maratha. The tribal small farmers have patchy land. There are landless families among the tribal as well as Maratha communities. The Maratha are aware of the fact that the benefits are not equal to all communities. In the discussions with them, they said that many farmers dug wells and made use of the increased underground water for the purpose of irrigation. As the tribal farmers lacked sufficient money to invest in digging the wells and putting the pipelines, they could not avail the benefits. This was because they had taken loans previously for different other purposes.

The landless people were not concerned with watershed development. They neither perceived the benefits nor the limitations. They earn their livelihood from labour work in nearby villages and by migration to outside places. They did not mention any change in their livelihood

Watershed development entails transfer of new technology and culture. It involves environmental understanding, technical problem solving, after care of the work carried out, following the three conditions laid down in the initial stages of implementation and maintenance. Overlook at these aspects is likely to lead to failure of the programme and consequently loss to the community. The people talked of neglect of after care of plantations done during watershed treatments in the village Kangar. The bans were not sincerely followed. The trees were not watered and the animals were left free to graze. So the number of trees that were actually planted was reduced. Even then the benefits are not less. In Kangar, the watershed work is under repair. Due to deposition of silt, the water retention capacity is reduced. During the watershed development period in Kangar, small dams (*tali*) were made in the village for water storage and percolation. However, due to technical reasons water percolation does not take place as desired causing water scarcity in the village in the months of April and May. Water is available for drinking all over the year but not adequate for irrigation. Thus, technical failures result in slowing down the pace of change.

In Mhaswandi, during the period of watershed development, there was a compulsory ban on grazing animals. A fine of Rs. 500 was imposed on those who did not follow the conditions of watershed development. Even after the completion of watershed development for more than 7–8 years, the rules are still being exercised and followed. Watershed factors in association with culture affect the livelihood patterns of the tribal people as discussed earlier. A lot of work needs to be done to minimise these effects in terms of social equity.

As observed by the community, with watershed development the purchase of Jersey cows has affected the milk consumption among children. The children do not like the taste of milk of Jersey cows.

Participatory Growth Monitoring of Children 0–5 Years and Relevant Social Change

Attitudinal Change

The women health promoters shared their experiences of monthly weight and height check up of children 0–5 years. They social with watershed development there has been a change in the attitude of the tribal people. Earlier the people refused to weigh their children due to evil eye notions (*Nazar lagate. Balachi tabyet kharab hote*). But they have started realising the importance of recording weights of children. They have realised that it is from weights that the health of the child can be judged. Whenever the mothers are called by the health promoters, they bring their children to the *Anganwadi* centre. The health promoter said that the mothers do not give excuses instead they ask for reasons if visit is not paid (*kahi karna sangat nahit ulat nahi gelo tar vichartat*). So there is a change in the attitude.

Awareness of Nutritional Status of Children

With participatory growth monitoring activity, the health promoters and the *Anganwadi* workers together pay home visits. The mothers are made attentive towards the nutritional status of their children. The health promoters encourage the mothers to plot the weight of their children on the growth chart. A growth chart developed by SAMPADA TRUST is provided to the health promoters which differs slightly from that of *Anganwadi*. By knowing the weight variation of their children every month, the parents come to know about the nutritional and health status of their children.

Attention to Diet of Children

As a part of growth monitoring activity, healthy baby competitions are organised once in a year in the village. Every mother wishes her child to come first and win the prize. The activity encourages women to pay more attention to the diet of the children. Observation indicates that mothers were keen about getting their child rewarded in healthy baby competition. They feel bad if their child misses a number in these competitions. A spirit of competition for proving the health of the child has been triggered. The tribal women were also found taking interest in healthy baby competition. They participate in the programme.

Change in Treatment Seeking Behaviour

The health promoters said that the people take their children to hospital immediately when they fall sick. If the weight decreases, the people relate it to improper food intake of the child. The Maratha perceive that a change in the treatment seeking behaviour of the tribal people has come about. Earlier when the children fell sick they would not take them to the doctor. Instead they would take them to their traditional healer locally called as *Bhagat* or *Devrushi*. A delay in seeking treatment led to the death of the children in certain cases. With inputs from growth monitoring, this practice has changed. The children are taken to the doctor for treatment.

Problem Management

The mothers are getting aware of the things to be done if the child is undernourished. A tribal woman having a malnourished girl informed the status to the researcher. She said that she feeds crushed groundnut and jaggery to the child to improve her weight. Crushed groundnut mixed with jaggery helps increase blood *(dane gul rakta vadhnyasathi)*, said the woman. The tribal women talked of the importance of drumstick leaves (*shewga cha pala)* in child nutrition.

The impacts of the two kinds of development inputs, watershed development and participatory growth monitoring as described in the text are summarised in the Figure 5.

Figure 5

Areas of Impacts of Watershed Development Inputs

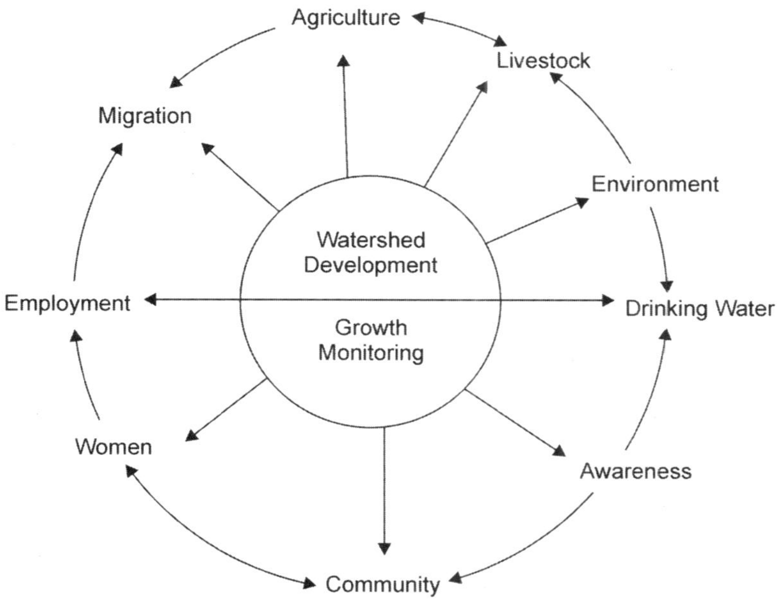

Discussion

Watershed development has definitely contributed to the increase in the average income of the people. The water conservation works carried out under the watershed development projects make the real difference. Water ensured gives assured returns, examples of crop assurance and enhanced crop yield. More water than before is available for drinking, for livestock, domestic use and agriculture.

Although the benefits to large farmers are more, the small and medium farmers, including the tribal small land owners have also benefited. The tribal people particularly were said to be experts in tomato cultivation by the non-tribal communities. They get employment to work on tomato farms. There is definitely a social change among the tribal community. Since time immemorial the tribal communities have been characterised by isolation, subsistence economy, their alienation from market forces. With increased crop production and sale of surplus as mentioned by the tribal people themselves, they are developing links with market. Watershed development as one of the forces affecting them is actively moulding their lives. It has facilitated the change. There is a change from subsistence to surplus economy among the tribal community. The pace and intensity of change when compared with their neighbours, the agricultural caste community Maratha, may be slow and uneven, but when a vertical and a horizontal comparison is made, a definite change is observed. Vertical comparison is comparing the present-day situation of the tribal people from the study villages with their past. Horizontal comparison is comparing the situation of the tribal people from the study villages with the tribal people from other regions at the same time. For a good judgement it is essential to make both types of comparisons.

The results obtained from data, viz., enhancements in personal hygiene, frequency of meals, and initiation of the process of reverse migration, mobility and exposure of women, community interactions through participation in village watershed committees, self-help groups and women committee, *Samyukta Mahila Samiti* are some of the factors contributing to social change as triggered by development inputs.

The impact of watershed development is likely to be uneven because of several reasons as follows:

1. Cultural variations between the communities.

2. Differences in the land related matters: land ownership, nature, soil quality, terrain, area of habitation. The caste Hindu Maratha are the primary beneficiaries. They are concentrated on the plains while the tribal Thakar and Bhil reside on the hilltop in the two villages. The inherent limitations of watershed crop in the data are in agreement with the case study by Kerr et al. (2002): "Watershed development projects are implemented from ridge to valley approach thereby benefits realised disproportionately downstream. By their nature, area development programmes offer benefits primarily to landowners, with landless people benefiting indirectly". The projects restrict access to resources that contribute to the livelihoods of the landless and this affects them worst. This holds true, for example, restrictions have caused sale of livestock like local cows and goats as free grazing is not allowed. This affects domestic milk production and consumption, petty income associated with it and consequently child nutrition and health particularly of the tribal community.
3. Dominant caste exercises control on village resources and insensitive towards livelihood issues of the poor.

With the rise in income levels, a commercial attitude is on the rise. Cash crops are the priority for the people. People concentrate on agriculture: sale of onion, tomato, potato and income from dairy. Many a times the mothers do not get adequate time to feed their children due to increased agricultural work. They are left with grandparents at home.

Although there is increase in awareness towards nutritional status, child health is yet to be perceived as a priority.

5

Measuring Culture Change: Concept Points Scale

Concept Points Scale Measuring Culture Change

Development organisations need to understand the process of change and its measure where development activities are initiated in a community. Understanding the process of change gives an idea of the kind of change that has taken place, how it looks like, the factors that have contributed into the making of change. Measuring change gives an idea as to what extent the change has taken place. It gives an idea of the effectiveness of a development programme and helps to improve upon the work.

During the study of understanding change in community perceptions across comparative groups, an attempt has been made to present qualitative data on culture change in the form of a scale. The scale helps not only to present but measure culture change. The chapter discusses the development of this scale.

The scale is called as "Concept Points Scale". The term 'concept' here refers to one's own perceptions. The scale is based on a number of responses given by study groups regarding the perceived factors affecting child nutrition and health. The scale can be called as "Perception Points Scale", "Health Concept Points Scale", "Health Perception Points Scale" to pinpoint the subject matter of change.

Theoretical Base: A Viewpoint

One's own perceptions/concepts are based on one's own or others' experiences, reading, listening to information, learning in schools or other educational centres, socialisation, exposure, may that be to the outside world or to the media, meeting different people, travelling to different places, own observation, self-reflection. Expression of perception is based on one's own organisation of thoughts, knowledge, judgement, experiences and confidence. The concept points measured on the scale depict the measure of concepts or knowledge a group possesses and is able to express in a discussion. Consistency of the concept points with scientific literature/practical life experiences is important in the analysis of change. A literature is a body of knowledge. Scientific literature means that kind of literature which is accepted as logical, accurate information by science made available in print and published forms. It could be that literature which is used for raising awareness in a community by a development organisation.

The following discussion will help to understand how the scale is and could be developed and how it can be used.

Methodology

The perceptions/concepts of the community regarding child nutrition and health were a part of the qualitative data. They have been gathered from three study groups/villages to understand the process of change a long time after the initiation of the watershed development and participatory growth monitoring inputs. The study groups are repeated here for clarity:

1. Watershed Development and Growth Monitoring Inputs Group: WSD+GM
2. Watershed Development Group: WSD

3. No Inputs Group: No Input.

Each study group is represented by a village with the kind of inputs initiated.

The objective of this part of qualitative data was to understand community perceptions regarding factors affecting child nutrition and health. The data have been collected from women, the mothers and the caregivers of children 0–5 years, other elderly women and men, from the community. For details regarding the methodology of the study, Chapter 2 may be referred.

The perceptions of the community health care providers group that is the *Anganwadi* workers have also been gathered and presented separately. The reason is that *Anganwadi* workers and the health promoters conducting growth monitoring as trained by Sampada Trust work in close collaboration with each other. A research question that emerged during the data collection was, "Do the *Anganwadi* workers too from the WSD+GN study group/village show a change in their perceptions because of collaborations with development workers?"

The analysis of the qualitative data which ran into several pages and books had been a very complex task consuming a lot of time and requiring a greater degree of concentration to find out whether change has really come about. A change was observed in the data but was not convincing. During analysis in the office, the collected data was referred to several times, telephonic conversations were held with the villagers wherever possible, field investigators were requested to get back to the field with a training to obtain specific information to confirm and reconfirm observations of change taking place in community. Understanding was achieved. The more complex task was to present the whole in a manner that would be simple to understand primarily from the point of view of development practitioners. The efforts on analysis led to developing a Concept Points Scale. The data on community perceived factors affecting child nutrition and health as observed across the study groups and across the ethnic groups were then numbered and numbers were plotted on the scale. Immediately the difference in health culture change between the study groups and between the ethnic groups was visible.

The application of the Concept Points Scale in measuring culture change is discussed with reference to its usage in our study. The details are described as follows.

The community concepts have been presented separately. Headings given to sections indicate the kind of data obtained from different groups to fulfil the objective. The data from both caste community Maratha and the tribal Thakar, Bhil and Lamani community are presented and discussed in comparison. The data from *Anganwadi* workers from the three study villages are also presented separately.

Applications of Concept Points Scale

The qualitative data are first presented. The information points from the data are numbered and put in a matrix and then a scale is prepared. The scale helps to show the rate of change across comparative groups, the communities and the health care providers group (*Anganwadi* workers) and how the groups are distanced.

Data of Village 1—Mhaswandi (WSD+GM Group)

Concept of Child Nutrition and Health among the Caste Maratha

The Maratha community from village Mhaswandi perceives personal hygiene/bodily cleanliness (*sharirachi vyavasthata)* as contributing to child health. A person who brushes his hands and legs maintains bodily cleanliness *(hat pay vyavasthit ghaslele)*. It is believed that a person who maintains hygiene, a condition when there is no dirt *(ghan)* on the body, never falls sick, meaning that the personal hygiene of the children should be so maintained. A child with cough and cold *(khokla aani sardi),* is considered as ill and therefore not healthy. Physical capacity *(sharirik shamata)* and freedom from illness *(aajar naslyavar)* and pain *(dukhat naslyavar) (nirogi)* were mentioned as other elements of the nutritional status and health of the child. Fitness as exhibited from these conditions gets reflected in one's physical appearance. A healthy person or a child looks charming *(takatak)* is the belief.

The women perceive food that is wholesome in nature *(yogya aahar, vyavasthit jevan)* as nutritious food. A bread made up of pearl millet (*bajri chi bhakri*) and a green leafy vegetable *(pale bhaaji)*, rice and dal, which is their regular food is considered as strength-giving. Their diet mainly includes bajra bread, vegetable made up of onion leaves or fenugreek, rice and dal made up of *tur*. Food as well as water are considered as important components of nutrition. They perceive that the consumption of food at specific timings and following a timely pattern adds to the nutrition. *Anganwadi* food is considered as containing all the nutritional elements required by the body *(jeevansatva)*.

They explained that such nutritious food is needed by the body to keep the children healthy. Nutrition is related to growth *(vaadh)* of the child and unless the child eats sufficiently enough as much as required by him, growth will not take place.

When asked about nutritional status of the children 0–5 years, the people mentioned weight for age of the child as the indicator. If a child is having very low weight for his age as is known from the *Anganwadi* then he is undernourished (*kuposhit, lahan: vayana jasta, pan vajanani kami*). The face and body of such a child looks very weak *(katkula)*. They perceived the cause of undernourishment as mother not consuming food, not taking injections and iron tablets regularly in her pregnancy. Such irregularity of the mother in taking food and medicines is because of her occupation with farm work.

The people believe that children's health can be judged on the basis of the *Anganwadi* records of number of malnourished (*kuposhit*) and diseased *(ajarpan)* cases in the village. As there are very few such cases in the village it can be said that the nutritional status of children is good. They said that health of the child can also be judged on the basis of records of the Sampada Trust who conduct growth monitoring of children 0–5 years in the village. It is believed that if a child is normal as per Sampada's record and growth gradation, the child is considered to be having a good nutritional status (*changla grade*) and healthy. As the growth chart of the NGO is huge and is put up on the walls of the *Samyukta Mahila Samiti (SMS)* office, a larger and a representative unit of all the women SHGs (self-help group) in the village, the record of children as per their grades is visible to the villagers.

They perceive that *Anganwadi* records the weight of the child from time to time to find out the low weight children *(vajan ghetat velchya veli)* and that the malnourished children are referred for special medical care.

The overall nutritional status of children is considered as good by the people. The performance is accredited by the people to the inputs provided by WOTR, Sampada Trust and work done by SMS. A Maratha elderly informant opined that the NGO pays great attention to the nutritional status of children (*te daksha aahet arogyachya babtit*). Earlier people would not give due attention to the health of the child. They never thought about the nutritional status or about vaccination of children. Earlier children suffered from measles and polio, which is not the case at the present. He believes there has been a change in the nutritional status of children 0–5 years because of the efforts of health promoters trained by Sampada Trust, the organisation working in the village. It is believed that their guidance and information to the mothers about the nutritional status of their children has brought about the real change.

When inquired about their ideas about nutritious food, the Maratha women of Mhaswandi listed out items made up of soyabean, sweet food items prepared from papaya, laddoos made up of groundnut and jaggery, items made up of drumstick vegetable. Women have made some of these recipes. They listed out the items from the information they obtained during the balanced diet demonstrations undertaken by Sampada Trust.

Perceived Illnesses and Causes among the Maratha

The Maratha women listed out the common illnesses and causes presented in the table as follows:

Table 5

Perceived Illnesses and Causes by Maratha

Common Illnesses	*Causes*
Cough and cold	Climate change
Fever (*taap*)	Climate change
Loose motions, diarrhoea (*patal sandas, julab*)	Not known
Gastro (gastro)	Excessive intake of sugar
Measles (gover) and chicken pox (*kanjinya*)	Not known
Chickengunya	Not known

Treatment Seeking Behaviour

In Mhaswandi, the treatment for children during illness is sought from private clinics and hospitals by the Maratha. The children are taken for treatment to taluka places, about 50 km away or nearby village, about 10 km away, having private clinics of their choice. Most of the Maratha have motor bikes and can have easy access to far off places for treatment.

Concept of Child Nutrition and Health among the Tribal Thakar Community

The Thakar of Mhaswandi judge the health of the child on the basis of his performance of the routine activities that are, eating of regular routine food, drinking milk, ability to digest.

The health of a small child is judged on the basis of its weight. It appears heavy (*Bal bharpur aahe*) if healthy. Availability of a number of food items to eat and drink besides daily food, rice and milk is considered as adding to the weight and nutritional status of the child *(khaya, pyaya bharpur aahe tar buva tabyetich haaye)*. Weight as an indicator of health occurred prominently in the discussion. For example, they mentioned that if a child is underweight, he should eat more whatever he can digest. Timely availability of food and breast feeding helps maintain the health of the child. Diarrhoea is mentioned as the sickness the children get affected due to lack of this. Although the sickness was perceived as related to the health status of the child in the discussion with them, it was not explicitly mentioned by the Thakar women. This could be probably because diarrhoea is a usual and routine kind of illness and freedom from diarrhoea was not thought of as a condition of the health status of the child. The role of immunisation in the prevention of illness to children was mentioned by women as contributing to the health of the child.

A pale body of the child *(suklela)* is considered as unhealthy. A sick child refuses to eat and drink milk.

The Thakar mentioned that their diet mainly includes *bajra* bread, vegetable/chutney *(Kalwan)* made up of *khurasani*, salt and chillies. They also have vegetable made up of onion leaves and fenugreek, rice and dal made up of *tur* and were not sure whether they were nutritionally rich items.

Regarding the nutritional status, a Thakar woman immediately mentioned that her daughter is malnourished *(kuposhit, kharab, walkuti)*. She said that she did not know the reason for malnourishment. Women mentioned that in cases when a lactating woman conceives, the growth of the second child born is better than the earlier one as it consumes all the breast milk and nothing is left for the previous or the first child. This has happened in the case of the malnourished child.

They mentioned that groundnuts and jaggery are nutritionally rich items. She gives the child crushed groundnut and jaggery for improvement in her weight and nutritional status *(Shengdane an gulani tabyet sudharte. Rakta vadhte)*. Groundnut is grown in the village and also by few Thakar families who possess land. Therefore, it is easily available to give to the child.

Perceived Illnesses and Causes among the Thakar

The Thakar mentioned about cough and cold, fever, diarrhoea as common illnesses among their children.

According to the Thakar, illnesses like loose motions are caused due to changes in water and food habits. They were not clear about the kind of changes in water. They get water from the tap. There is a water pipeline in their hamlet. They mentioned that consumption of sour fruits like tamarind, *bor* (Ber) leads to cough and cold in children.

Treatment Seeking Behaviour

There is a herbal medicine man among the Thakar. People prefer taking treatment from the medicine man and from the primary health centre.

It was observed that no medical treatment was sought by the Thakar in one case where a child was suffering from diarrhoea. Probably they must have thought that it will stop naturally. Home treatment is followed to cure cough. Cough is cured by feeding honey to the child 2–3 times in a day. The health promoters opined that the Thakar do not take their children to doctors immediately when they fall sick.

There are Government Primary Health Centres (PHC), one located at Bota and another at Aalephata, both at a distance of about 10 km. Aalephata is a place closer to the hilly terrain of the village where most of the Thakar live. They believe that small children are treated at PHC (*lahan mula mhatala mhanje sarakari madhyech bolavitat*). Therefore, they prefer PHC for treatment of children. They go walking to these places.

Concept of Child Nutrition and Health of *Anganwadi* Workers from Mhaswandi

The work of *Anganwadi* workers is primarily related to the area of child nutrition. There is one *Anganwadi* centre in the village.

The *Anganwadi* workers of Mhaswandi talked of a wide array of factors contributing to child nutrition status. They mentioned weight, height, socio-economic status, availability of irrigation in farms *(Panyakhalchi sheti)*, immunisation, timely feeding, awareness of nutritious food, freedom from disease, seeking treatment in time, inputs by *Anganwadi*, settled life and no migration as factors contributing to the nutritional status and consequently the health of the child. They were of the opinion that people having irrigated farms are well off and have the necessary cash at their disposal to purchase food items, to seek immediate treatment, that such people with a better socio-economic status send their children to *Anganwadi* and thereby benefit their own health. The *Anganwadi* worker was concerned about the inputs given by her. She expressed that from the weight of the child, she comes to know if protovita powder provided by *Anganwadi* for enhancement of nutritional status is fed to the child. She rated the nutritional status of the children in the village as good.

As per *Anganwadi* records, there is one case of malnourished (*Kuposhit*) child in the fourth grade among the Maratha. According to the *Anganwadi* worker, the cause of malnourishment was because the mother of the child was malnourished too. Her age at marriage was below 18 years having minimal spacing between two children.

According to the *Anganwadi* worker, low age at marriage, poor nutritional status of the woman and minimal spacing between the two children has led to malnourishment in the case mentioned.

Perceptions of Illnesses and Causes and Treatment Seeking Behaviour by *Anganwadi* Workers

The *Anganwadi* workers mentioned about cough and cold and the presence of worms *(janta)* in children 0–5 years. Change in weather conditions caused cough and cold and she attributed the cause of worms to unsafe drinking water *(dushit pani)*.

Cough and cold is believed as normal and no specific treatment is given to children, said the *Anganwadi* worker. The *Anganwadi* provides treatment for worms. A specific dosage of medicine is given to *Anganwadi* children every six months.

Data of Village 2—Kangar (WSD Group)

Concept of Child Nutrition and Health Among the Maratha

The Maratha from Kangar highlighted *Anganwadi* as having an important role to play in the health of the child. This is because body weight is measured in *Anganwadi*. There are four *Anganwadi* centres in the village. The women perceive birth weight and child health condition during birth as contributing to the nutrition and health status of the child. The women mentioned birth weight as also indicating the nutritional and consequently the health status of the children. The children with poor health conditions are advised by the doctor for intensive care *(kachechya petit thevtat)*. It is believed that such children often fall sick. A child with lower birth weight, that is below 2.5 kg is considered as malnourished *(kharab)*. They have no idea about the reasons for such conditions and measures for improvement of the health of such children.

Activeness of the child is considered as an important sign of good health of the child. The people mentioned that a child having poor nutritional status can be judged on the basis of his appearance. Weakness in appearance, in muscle accumulation *(ashakta, tabyet kami)* and body weight *(vajan kami aaheba)* are some of the signs recognised by the Maratha of Kangar as indicative of poor nutritional status. Performance of regular activities by children like *Anganwadi* attendance, studying, playfulness, singing songs taught in *Anganwadi,* engaging in conversations amongst them indicates that they are maintaining good health.

The comments of women of Kangar in a focus group discussion towards the mother whose three children had poor nutritional status speak about their perception. The women commented that consumption of pulses *(kaddhanya)* in diet will help in improvement of the nutrition component of diet of the child consequently improving the health status. Besides they commented that health tonics prescribed by doctor will help improve the health status of the children.

As *Anganwadi* is the major source working to improve the nutritional status and providing information about weight of the child, role of *Anganwadi* in nutritional status of the child is well-recognised as it was re-emphasised in conversations, the focus group discussions and in-depth interviews. Food frequency and the availability of food as per the requirement and appetite of the child are recognised by the Maratha as contributing to the enhancement of his/her nutritional status. They mentioned that there is no shortage of food and food is available all year round as wanted by the children among the Maratha. Therefore, there can be no problem with the nutritional status of children.

Perceived Illnesses and Causes by Maratha

It was found that the community women and men and the *Anganwadi* workers from Kangar hold the same perceptions on illnesses. Therefore, they are presented together in the following table:

Table 6

Perceived Illnesses by Maratha and Anganwadi Workers, Kangar

S. No.	*Common Illnesses*
1.	Cough and cold (*sardi, khokla*)
2.	Fever (*taap*)
3.	Gastro (gastro)
4.	Diarrhoea (*julab*)
5.	Vomiting (*vantya*)
6.	Scabies (*kharuj*)

The people of Kangar listed out but did not say anything about the causes of the illnesses. The community women as well as the *Anganwadi* workers mentioned about climatic change as the reason, for illnesses like diarrhoea, vomiting and scabies but did not give reasons for other illnesses. The *Anganwadi* worker stressed nutritional status as the affecting factor.

Treatment Seeking Behaviour

During illness, the Maratha of Kangar villages seek treatment from a private doctor, a child specialist located at taluka place, about 12 km from the village. Dissatisfaction about government services like improper attention paid by the PHC staff as perceived by the villagers makes them seek treatment from private clinics. In case of no improvement in the situation then they resort to the local healer called *devrushi*. In case of problems like fears among children, getting up from sleep in the night due to fear the children are taken to the traditional magico-religious healer locally called *devrushi.* The child is treated with a magico-religious treatment by giving ash (*angara*) as medicine and putting a thread around the neck (*dori ghaltyat*) of the child.

Concept of Child Nutrition and Health among the Tribal Bhils

Freedom from disease, timely feeding of child were considered as contributing to the nutritional and health status of the child. A cheerful, smiling and active child, who looks charming is considered as good in health (*nirogi, tyala ajar nahi pahije, hasta khelta rahila pahije, jevan velevar dile pahije, tavtavit disto*). It is also perceived that the nutritional status of the child is dependant on his body structure *(shareer prakruti).*

A child whose health and nutritional status is poor is always sick. He is prone to Malaria. He has cold, running nose, fever, diarrhoea. The health status can be recognised from his appearance. A sick child keeps crying, his face looks dry (*tabyet kharab hote tyachyawali, ajar hotyat, tap yeti, julab hotya, sardi hoti, chehara sukel rahato, sarkha radto*), said the women. A Bhil tribal informant mentioned that rickets *(muddus)* was common among the Bhil a few years back. Now no such illness is observed from the past four to five years.

A change in the source of drinking water causes diarrhoea and fever. Weather changes also lead to the illnesses among children *(pani badalta. Vihiricha rahata kadhi kadhi nalacha*).

The men and women mentioned that their diet usually includes *bajra* bread and chilli *(bhakar an mirchi)*. The chilli is crushed and made in the form of chutney. They said that it is the only food they have and do not eat anything else. Dal made up of *tur/urad* or gravy made up of potato (Kalwan) is also there with the pearl millet bread.

Researcher's Comments: The statement made by the Bhil informant from village Kangar about rickets is valid. No case of rickets was found in the village by a doctor associated with WOTR who conducted clinical examination in Kangar and other villages.

Perceived Illnesses by Bhils

Fever, diarrhoea, cough and cold, vomiting are the common illnesses among the Bhil children as mentioned by the people and the *Anganwadi* workers. There are two cases of malnutrition among the Bhil, a boy and a girl, of age four years. They did not say anything about the causes of the illnesses.

Treatment Seeking Behaviour

During illness, treatment from Government Primary Health Sub-Centre is usually preferred. The medicines are given by the nurse there. However, if the child is serious he/she is taken to the Rahuri Sugar Factory hospital. If the child doesn't respond to the doctor's treatment he/she is taken to traditional healers (*dev dev asa kela ki tyani utartya amchi pora, jadu tona mantyat na tyala*). There is an old traditional healer living in a nearby village, called *devrushi*. Most of the village people of all communities seek treatment from the local healer. He provides a magico-religious treatment by putting a thread around the neck of the patient (*dorya taktat galya madhe*).

Perception of *Anganwadi* Workers from Kangar

There are four *Anganwadi* centres in the village.

In-depth interviews have been conducted with *Anganwadi* workers of two centres.

The *Anganwadi* workers are of the opinion that it is the body weight that reveals the nutritional and health status of the child. Malnourishment in children is primarily because of the neglect by the mother. Timely feeding by mother helps maintain the nutritional and health status of the child which is known from body weight. A decrease in body weight indicates neglect of the mother. Poor nutritional status of the parents also contributes to malnourishment in children.

Regarding the signs and symptoms of nutritional status of the children, she mentioned activeness, playfulness, cheerfulness as indicating good nutrition and health condition of the child. A malnourished child has dry face and body *(jhuralya distat cheharyavar, angavar)*. They said dryness of skin is a typical feature of malnutrition and that a healthy baby has a charming face *(tawtawit chehara)*.

Data of Village 3—Kalas Pimpri (No Input Group)

Concept of Child Nutrition and Health among the Maratha

The Maratha women from Kalas Pimpri expressed frequency of food consumption that is consumption of four to five meals a day is considered as contributing to nutritional and health status. Body colour, specifically fairness of skin and appearance are considered as indicators of child health. Black colour, thin and pale looking body is considered as indicating poor health status (*kalsaar nahi ani suklya sarkha nahi paahije*). Shortage of food, that is, meals less than four to five times is considered as affecting health. Freedom from disease and body weight were not explicitly mentioned as contributing factors for health but stated their children's illness as a factor not keeping them healthy. Illnesses are perceived as causing weight loss affecting health *(khokla, sardi jhali ki vajan vadhat nahi)*.

The Maratha women of Kalas Pimpri village perceive as having regular diet (*bajra* bread, vegetable) at frequent intervals as ideal for providing nutrition to the body and consequently contributing to the health of the individual. The intervals would be as follows said they: First meal in the morning after getting up, then breakfast, lunch, evening meal and dinner at night.

The key informant interviews at Kalas Pimpri highlighted the literacy level, working conditions of mother, migration for employment as causative factors for the neglect of children consequently affecting their nutritional status. One of the key informants shared his experiences. Few children of sugarcane factory labourers were malnourished. They were identified and sent for treatment by the health workers to the government rural hospital. Treatment with diet, medicines and child care for fifteen days resulted in improvement in the nutritional status of the children.

The key informant interview at Kalas Pimpri highlighted body composition (*mulancha sharirik rachnevarun manta yeil*) as indicative of nutritional status of the child. A malnourished (*kuposhit*) child shows no body growth, belly out, no freshness in skin, appearing unhealthy, lack of facial glow (*vaad na hone, pot pudhe yene, tvacha phar upyukta vatat nahi, chehara varcha tejaswipana rahat nahi, haat pai phar jirn awasthet astat*). Facial glow, freshness on cheeks and a smile are considered as indicative of good health of the child (*tavtavitpana galavar rahane kiva tavtavit hasne*).

Perceived Illnesses and Causes

The following table presents the perceived illnesses and causes by the community and the *Anganwadi* workers.

Table 7

Perceived Illnesses by Community and Anganwadi Workers, Kalas Pimpri

Common Illnesses
Cough and cold (*sardi, khokla*)
Fever (*taap*)
Loose motions (*patal sandas*)
Vommitting (*vantya*)

There was a case of a girl child having two holes in the heart because of which she did not keep healthy and was underweight as mentioned by the mother of the child weight less than 10 kg at the age of 18 months. The Maratha women from Kalas Pimpri said that they did not know the reasons for illnesses of their children.

The *Anganwadi* worker opined that the underweight children were prone to illnesses.

Treatment Seeking Behaviour

In Kalas Pimpri, there is no doctor in the village. In case of emergencies or at night time if a vehicle is needed, it is called by making a phone call to the drivers staying in nearby areas. The people seek treatment from private clinics and hospitals at places like Koradgaon, about 5 km away, Shevgaon, about 20 km away and Pathardi, a taluka place, about 25 km away. The doctor at Pathardi is considered as good for children. They go to these places by personal vehicles—motor bikes or by S.T. bus. There is a government Primary Health Centre (PHC) at Koradgaon and at Pathardi. The people do not like to seek treatment from the PHCs because they feel that the PHC staff does not pay proper attention, the people feel neglected as they do not give medicines and injection, charge for the health services and make the patients visit the hospital repeatedly.

Concept of Child Nutrition and Health among the Tribal Lamani

The Lamani mentioned weight as the indicator of a child's health. Low weight babies are malnourished. Illness causes weight loss *(samja dukhaya aala tar kami vajan aala. Changli tabyet rahili tar vadhun rahata)*. In the age group of 0–5, weight fluctuations are considered as natural because of growth parameters like emerging of teeth. It is believed that the stability in the health of the child comes as the child grows old *(te motha jhalyavarach tyachi tabyet changli rahate)*. They believe that the economic status of the family determines the availability of nutrients the child gets *(jyachya tyachya paristhiti nusar jeevansatva bhetatat)*.

The Lamani said that *bajra* bread is nutritious food *(bajrichi bhakri dankun)*. It is filling and satisfying. They purchase wheat on ration. But chapattis are not adequate *(aapla khatala sakat, chapatti kutha purati?)*

The signs like crying, refusal to eat and drink, sleepiness, inactiveness (not playing) mean ill-health conditions in children. Crying leads to loss of energy affecting the health status of the child.

Perceived Illnesses and Causes by the Lamani

The Lamani mentioned fever, cough, cold, vomiting, diarrhoea, chicken pox *(kanjinya)* as common among their children.

Vommitting and diarrhoea are considered as natural and common among children as they occur during the growth stages of the child *(He aata jar lahan mulga asla tar tyache dat nighatat, tyala julab hotat, tyala ulti hotat, tyachyamule ti tabyet kami hote. Te motha jhalyavarach tyachi tabyet changli rahate)*.

Fever, cough and cold are believed to be caused due to weather changes. They said that children cannot tolerate weather changes, that is why they fall sick. Hot and cold climate fluctuations are common. They said that it is very hot in villages and is not like the houses in cities. The tin roofs of the houses in villages generate more heat leading to illnesses like fever. Chicken pox is perceived as some sort of illness affecting god *(devala dole yetat)*. Children eat mud and get affected with such illnesses.

Treatment Seeking Behaviour

Among the Lamani, the children are taken to private clinic at Koradgaon during illness. The village Koradgaon is about 5 km from the village. The PHC is at Pathardi. Pathardi is far off from the village Kalas Pimpri and the travel expenses are more than the services obtained at Koradgaon. At Koradgaon, there is a PHC but people do not go there because the doctors charge for health services.

Perception of *Anganwadi* Workers from Kalas Pimpri

The *Anganwadi* worker opined that weight as per age was the only criteria that determined the nutritional status of the child. Children who were underweight as per the growth chart of *Anganwadi* were considered as having poor nutritional status affecting the health of the child. She opined that majority of children in Kalas Pimpri village, irrespective of caste and tribe belonged to grade II nutritional status (*sthiti madhyam swarupachi aahe)* as per *Anganwadi* records. Late weaning and insufficient quantity of supplementary food (*saha mahine te varsha chya mulanna varcha aahar pahije asto manje chalu karaycha asto. Pan khedegavat manje yevdha varcha aahar dila jat nahi*) were pointed out as the factors responsible for the situation.

The nutritional status of girl child is lower than that of boys. However, in *Anganwadi* worker's opinion it is not due to gender differences as perceived by the community but it is dependent on the family's financial status.

Perceived Illness Causes and Comments on Treatment Seeking Behaviour.

The *Anganwadi* worker too mentioned about illnesses like cough and cold, fever, gastro as common among children 0–5 years. Nutritional status was attributed as the causative factor for sickness. She said that the PHC nurse identifies the sick and gives medicines during her visit in the village.

Data Analysis

The data of three villages have been analysed and the comparative analysis is presented in the matrices below. A comparative analysis has been made across the communities. This sort of comparison is essential to understand the pace and process of change occurring in the two major communities, a dominant caste and the tribal groups, both at two different ends of change and development, giving a holistic picture of the process of social and cultural change in the study villages.

Matrices showcasing the factors affecting nutritional and health status of children 0–5 years of three villages have been presented.

Table 8: Showcases concepts of Maratha communities.

Table 9: Concepts of Tribal communities.

Table 10: Concepts of *Anganwadi* workers.

Table 8

Factors Affecting Child Nutrition and Health: Concepts of Maratha Community in the Three Villages

	Comparative Groups		
	Mhaswandi (WSD+GM)	*Kangar (WSD)*	*Kalas Pimpri (No Input)*
1.	Personal hygiene	——No mention of personal hygiene——	
2.	No mention of body weight	Body weight	Body weight
3.	Physical capacity	Birth weight	No mention of capacity, birth weight
4.	Freedom from illness and pain	No mention of freedom from illness and pain	Freedom from disease, illness
5.	Constant cough and cold leads to poor health status	——No mention of illness——	
6.	Role of *Anganwadi:* Food, growth recording	Role of *Anganwadi:* recording of weight	No mention of role of *Anganwadi*
7.	Role of NGO in growth monitoring	——No NGO is working on growth monitoring——	
8.	Wholesome nature of food	Consumption of pulses	No mention of nutritious food
9.	Food and water as important components of nutrition	Importance of health tonics as a component for child nutrition	No mention of nutrition components
10.	Relationship of food sufficiency with growth	Availability of food	Availability of food: Food shortage
11.	Disciplined quantity of food	Food frequency	Food frequency
12.	Timely pattern of food consumption.	——No mention of timely pattern, relationship of nutrition with growth and role of immunisation——	
13.	Relationship of nutrition with growth		
14.	Awareness regarding immunisation		
15.	Visual impact of growth chart	——Visual impact of growth chart is absent because there is no growth monitoring by NGO——	
16.	Weather change	——No mention of weather change——	

Cont'd...

Cont'd...

17.	Relationship of child health with mother's health: pregnancy care, food intake	No mention of relationship of child's health with mother's health	Relationship of child's health with mother's socio-economic status: literacy and occupation
18.	Listed out varied nutritious food items	Regular diet considered as the only available and nutritious food	No mention of nutritious food items
19.	——No mention of migration——		Migration

Table 9

Factors Affecting Child Nutrition and Health: Concepts of Tribal Community in Three Villages

	Comparative Groups		
	Mhaswandi (WSD+GM)	*Kangar (WSD)*	*Kalas Pimpri (No Input)*
1.	Digestion capacity	No mention of digestion capacity	No mention of digestion capacity and illness/disease
2.	No mention of freedom from disease as influencing child health	Freedom from disease/illness	
3.	Body weight	Body structure	Body weight
4.	Timely food intake and feeding	Timely food intake	No mention of food intake
5.	Availability of varied items	No mention of availability of food items and economic status	Economic status: availability of nutritious food
6.	Conception during lactation	——No mention of influence of conception during lactation——	
7.	Immunisation	——No mention of immunisation——	

Researcher's Comments on Data Related to Signs and Symptoms of Child Nutrition and Health Status

Regarding the signs and symptoms of child nutrition and health status, particularly of malnourished children mentioned by the community in three different villages, it is found that the WSD+GM group mentioned few signs than those from the other two villages. The information that they possess outnumbers signs and symptoms.

Table 10

Factors Affecting Child Nutrition and Health: Concepts of Anganwadi Workers in Three Villages/Comparative Study Groups

	Comparative Groups		
	Mhaswandi	*Kangar*	*Kalas Pimpri*
1.	Weight	Weight	Weight
2.	Height	——No mention of height——	
3.	Socio-economic status (SES): education, financial status of family	No such mention of SES or financial status	Financial status
4.	Irrigation	——No mention of irrigation, immunisation——	
5.	Immunisation		
6.	Timely feeding	Timely feeding: role of mother important	——No mention of timely feeding——
7.	Settled life and no migration for employment	——No mention of migration——	
8.	Treatment seeking in time	——No mention of treatment seeking——	
9.	Relationship of child health with mother's health: Nutritional status of mother, age at marriage and spacing between two children	Relationship of child health with mother's health: Nutritional status of mother	No mention of mother and child health relationship

Note: A specification of no mention of a particular point in the table means no relevant point or point of similar concept has been expressed by the group as compared to the other groups. It was specified in the table so as to sort out the perception points and make statements in comparison.

The community difference also seems to be prominent. The caste community Maratha exhibit more information and mention few points about signs and symptoms than the tribal communities. The Maratha from WSD+GM group are particularly more conscious. They seem to be judging the nutritional and health status of the child on the basis of the information/knowledge they possess rather than the appearance of the child.

Regarding the signs and symptoms of nutrition and health, the tribal people from Kalas Pimpri have mentioned more number of points, while the tribal communities from two other villages have mentioned almost equal points. The tribal people from all three villages judge the nutritional and health status of children on the basis of both the information/knowledge they possess and the signs and symptoms of poor health.

Discussion

Change within Communities

Caste: The Maratha from WSD+GM group gave a wide array of information than other study groups. Their perceptions include components broadly related to personal hygiene, food, nutrition, growth, child health, mother's health, illness, role of *Anganwadi*. It is pertinent to note that they do not speak anything related to weight: birth weight and body weight of the child in an area where growth inputs are given.

The Maratha from WSD group perceive components related to weight: body weight and birth weight as a separate point, food, health tonics.

The Maratha from No Input group perceive factors related to food: shortage is expressed as a factor, body weight, illness and migration. It is pertinent to note that migration is not mentioned by the other two groups. Migration as a factor affecting child nutrition and health is a practical life experience which is probably less in existence in the other two study groups, that is, the Input groups.

Tribe: The tribal Thakar from WSD+GM group perceive factors related to food, child health and mother's health. They have correlated the concept of health primarily with weight of the child. Weight indicates the health of the child has been expressed by the Thakar women. Availability of varied food items has been of prominent concern for them in contributing towards child health. When compared with their counterparts, the caste Maratha, this point was not expressed by them which could be because they have been prosperous with food and milk as evident from their background and such is not the case with the Thakar as evident from their varied means of livelihood. Probably the difference is because of this variation. The women are aware of their children's nutritional status. A Thakar woman herself mentioned in the focus group discussion that her child was malnourished. She used the term *kuposhit* to explain her child's status. She said that she got to know about her child's nutritional status from the health promoters. The health promoters and the *Anganwadi* workers work in close coordination with each other. They pay home visits together for growth monitoring of children. If one of them weighs the child, the other worker notes it down. The four women help each other in taking measurements, noting it down and explaining the status to the parents.

The tribal Bhil from WSD group perceive factors broadly related to illness, physique of child and food.

The Lamani from No Input group perceive factors broadly related to food and weight. For them economic status determines the availability of nutritious food. They stressed on weight as the indicator of nutritional status of the children affecting the health of the child. Like the Thakar they do not give a comprehensive list of causative factors.

Observation indicates that most of the tribal people including all communities do not seek treatment in the initial stages of the illness. As the illness gets prominent, they seek treatment from the local healer or the PHC. A prolonged illness is believed to be due to supernatural reasons. Hence, they seek treatment from the local healer. In case it gets severe, they then take the child to taluka level government hospital, the rural hospital. The traditional and modern remedies are sought almost simultaneously.

Health Care Providers

The *Anganwadi* workers from WSD+GM group expressed a wide array of information points broadly related to child health, relationship with mother's health, agriculture, socio-economic status, food intake. Height has been expressed by none except this group.

The *Anganwadi* workers from WSD group talk of weight and mother as important figure in child health: her role in feeding, her nutritional status.

The *Anganwadi* workers from No Input group perceive factors related to weight and the financial status of the family.

The role of body weight, and appropriate food intake in the maintenance and enhancement of the nutritional status of the children is recognised irrespective of the development inputs. The matrix also shows that the availability of food is of concern to the tribal community irrespective of the watershed development activity initiated in the area. Change although has crept in, is comparatively slow. The reason could be that the tribal community has inherent problems related to land ownership, settlement in relatively inaccessible forest and hilly areas, isolation from the mainstream due to which they cannot keep pace with the development achieved by the other dominant mainstream communities.

In Mhaswandi, there is one *Anganwadi* centre. The health providers group is large, with health promoters in addition to *Anganwadi* workers there are more professional people available to guide and pay regular home visits. As and when they come across any illness in a house, they advise the people to seek immediate health care. Even if the children staying far off are not able to attend the *Anganwadi,* the regular visits and inputs of the health care providers group could be of some help in contributing to child health care. The situation of village Kangar is slightly different. There are four *Anganwadi* centres. The village has dispersed settlements while in village Kalas Pimpri, there is one *Anganwadi* centre.

The *Anganwadi* workers are occupied with their routine work, maintenance and completion of records, meetings, reporting, coordination with the government health staff and helping the visitors in the village. The *Anganwadi* workers are doing commendable work with the entire work load. The people realise the role of *Anganwadi*, understand the relationship of body weight with nutritional and health status of the child wherein *Anganwadi* has played a major role. However, due to busy schedule, it is likely that they have little time left to frequently interact with the parents and advise for child care.

Treatment Seeking Behaviour

Most of the people prefer to seek treatment from private clinics and hospitals. Most Maratha from Mhaswandi have their own motor bikes and travel even up to 50 km for treatment. The Maratha from Mhaswandi are rich farmers earning huge profits. They spend on treatment in private sector by travelling large distances.

In Kangar, the Maratha seek treatment from private as well as a local traditional healer, *devrushi*.

The Maratha and the Lamani tribal people from Kalas Pimpri seek treatment from private sector. They spend money to go to far off places for the purpose of treatment. Those who don't have their own vehicle rent a four wheeler vehicle in case of emergencies and travel by local buses if not so urgent.

The Thakar from Mhaswandi and the Bhil from Kangar seek treatment from government Primary Health Centre (PHC). They believe that PHC is for children's treatment. That the children are called at PHC (*mula mhatli ki tyanna tikdach bolvityat na*).

Development of a Tool: Concept Points Scale

The data analysis presented in the form of matrices has helped in the development of concept points scale to understand and measure change.

Looking at the three matrices, a scale of concept points has been developed. Concept points indicate the knowledge possessed by the groups. The knowledge expressed as perceptions exhibits a great difference and a change as seen from the matrices. The difference is noted in terms of the number of concept points mentioned. More number of factors affecting child health have been expressed by the WSD+GM group.

Allotment of Scores

A score of one is assigned to each point expressed in perception. No scores are given to the points where the role of NGO in growth monitoring and visual impact of growth charts are perceived as factors enhancing nutritional status of children because these impacts are absent in other two study groups. A group or community who does not mention a relevant point mentioned by the other groups, get zero points.

The status of scores of the comparative study groups and the communities is as follows:

Table 11

Points Scored by Study Groups and Communities

Comparative Groups	*Concept Points Scale*	
	Caste Community	*Tribal Community*
WSD+GM	15	6
WSD	8	3
No Input	6	2

Figure 6

Scale Depicting Change in Communities from Study Groups

Caste Community

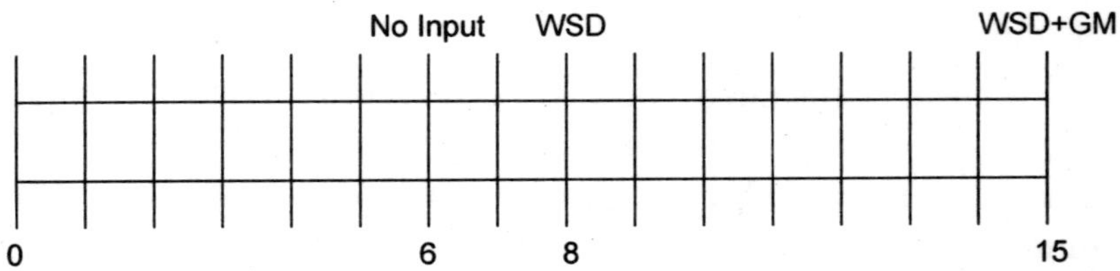

Tribal Community

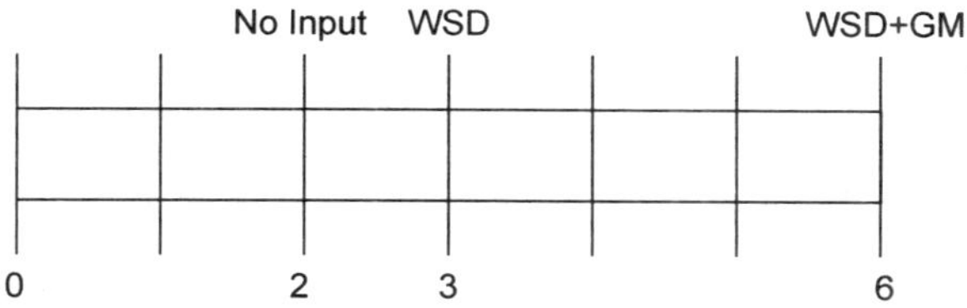

The caste as well as the tribal communities from WSD+GM group score higher than the other two groups. This indicates culture change in WSD+GM group as against the other two groups. The score matrix also points out that there is less/negligible difference between the WSD and No Input groups which means that participatory growth monitoring is playing a significant role than the only input of watershed development in terms of health culture change. The list is comparatively lengthy in the first group.

It implies that the people have expressed more ideas in the first group (WSD+GM) and stand slightly apart from the other two groups (WSD and No Input). The pace of social change in the first group appears to be slightly faster than the other two. The expression of ideas in a short duration of discussion also points out a variation from the other two groups.

Contentwise too, the information is more in the WSD+GM group. Contentwise, in the village Mhaswandi, the WSD+GM group exhibits rich and extensive information. None of the informants speak of food shortage or availability of food as a factor influencing child nutrition. Food shortage is particularly mentioned by the village Kalas Pimpri, a No Input group. Availability of food and the problem of food shortage are uppermost in the minds of the people of Kangar and Kalas Pimpri respectively. The tribal people from Mhaswandi talk of the availability of varied food items as against the other two villages. The tribal people from No Input study group say that economic status determines the availability of food items.

The content of the perceptions speaks a lot of the reality and the development processes in place. However, due to time constraints not much has been analysed from the point of view of what the contents speak.

Judged on the basis of the number of points expressed and the content of information, the WSD+GM group reveals more of health consciousness.

Another important finding that the scale points out is the existence of a large gap between the caste and the tribal community in any comparative group. Although the rate of change of the tribal community in WSD+GM group is faster than those from WSD and No Input group, the overall rate of change is slow among the tribal communities as compared to the caste community. The development inputs need to direct their efforts to the tribal community for better impacts.

The study highlights that watershed development inputs are causing the desired impact on the population. To make them more effective, development efforts should pay greater attention on the following two areas:

1. Watershed development group which means integrated and participatory watershed development needs to be made more effective and
2. Attention to the disadvantaged section of the village is more important to make watershed development inputs equitable.

Table 12

Concept Points of Anganwadi Workers Group

Comparative Groups	*Concept Points of Health Providers Group (Anganwadi)*
WSD+GM	9
WSD	3
No Input	2

Figure 7

Scale Depicting Change in Anganwadi Workers from Study Groups

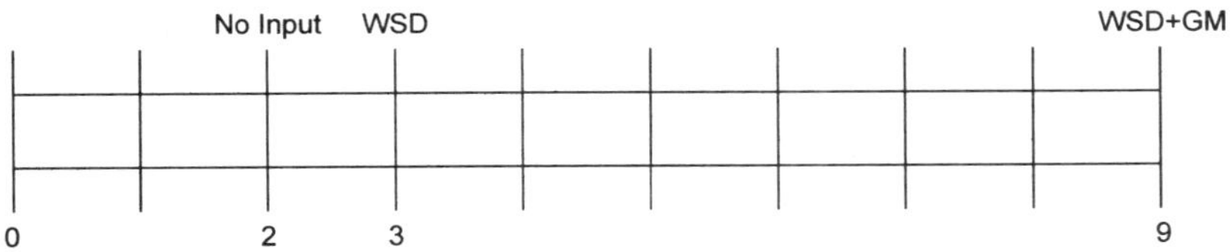

When the awareness levels of *Anganwadi* workers from the three study groups are compared a difference is found. Even at the level of *Anganwadi* workers, the WSD+GM group exhibits more information than WSD and No Input groups.

Comparing Community Consciousness with Health Care Providers Group

A very important fact that comes to light is when a comparison of *Anganwadi* workers from WSD+GM group is made with the caste community Maratha from the same group, the caste community Maratha score more than *Anganwadi* workers reflecting the impact of participatory growth monitoring input in watershed village. The Maratha from WSD and No Input group too are much ahead of *Anganwadi* workers from their respective groups. All this denoted greater health consciousness of community as compared to the health care providers (*Anganwadi* workers). The health consciousness of *Anganwadi* workers is almost near to the tribal community which denotes that both are on one scale of change.

This denotes that upgradation of knowledge of *Anganwadi* workers is drastically needed.

The *Anganwadi* workers with no growth inputs in their villages are behind the tribal community from WSD+GM input group.

The numbers of the scale speak of the rate of change. Greater the number, greater is the rate of change. It is seen that when watershed development is coupled with participatory growth monitoring inputs, it gives superior results in terms of development. The impact of participatory growth monitoring is superior and it is a significant input that should be incorporated as a component of integrated watershed development.

The concept points scale catches the change and can be used with regard to any community or group.

It is realised here that no development input works in isolation. The same applies to watershed development and growth monitoring inputs. Social change is a result of many different factors. Education is one of the important factors. A change occurs even without development inputs. However, the rate of change is faster with development inputs.

Factors like watershed development inputs, women's institutions, collaboration of NGO workers (watershed development and participatory growth monitoring NGOs) with the government *Anganwadi* workers, the visual impact of the growth charts, the components of growth monitoring, community participation are some of the key features that have proven to be important inputs leading to social and health change.

The concept points scale for three groups, viz., the caste community, the tribal community and the *Anganwadi* workers is shown as follows:

Scale Depicting Change

The study highlights some significant findings as follows which need a closer analysis:

1. The village with watershed development and growth monitoring inputs shows a higher impact, as observed in both caste and the tribal communities and the health care providers group (*Anganwadi* workers).
2. There is a definite change observed in the tribal communities.
3. The rate of change is accelerated in the caste community and is slowed down in the tribal community when the two communities are seen in comparison.

4. There is accelerated rate of health culture change found in communities vis-à-vis the *Anganwadi* workers. The *Anganwadi* workers with no growth inputs in their villages lag behind the tribal community who are given growth inputs.

The differences in the rate of change resulting from development inputs in the two different communities are possibly due to the differences in the cultural characteristics of the two ethnic groups, the caste and the tribe. If we look at the culture of the two communities, we may find an explanation for the difference in change. A thought regarding this has been put forward here.

Culture and Scale of Change

Caste Community: Maratha

The Maratha is an agricultural caste community, the land owning and land tilling caste possessing cattle (Karve, 1991). The Maratha have predominant population in all villages of Ahmednagar. They are the primary beneficiaries of watershed development due to land ownership in Mhaswandi as well as in Kangar. Some of them have large land holdings. In villages where watershed development projects have been initiated, the rainfed lands of the Maratha have been turned into irrigated lands. Floriculture, cash crops and dairy are the sources which have earned huge profits to the Maratha. The village Mhaswandi has 23 self-help groups and a women's committee, 18 of which are exclusively of Maratha women. Maratha women are more in number representing these institutions may be because the population of Maratha is larger. The SHG women are involved in a number of activities related to child nutrition. The mothers of children 0–5 years, members of SHG are involved in preparing *Anganwadi* food in rotation. The SHG women are involved in organising the healthy baby competition in the village. The participation of women in *Anganwadi* programme being a part of SHG, their concept of child health definitely exhibits a greater impact of participatory growth monitoring inputs initiated with watershed development at background besides other influences.

Tribal Communities: Thakar and Bhil

On the contrary, the tribal community is known to be the community residing in forests and hills. Their way of life, traditions, belief system, knowledge, regions that they occupy shows a close relationship with forest and nature. The tribal communities Thakar and Bhil from Mhaswandi and Kangar respectively are settled communities. They possess less land in the village as compared to the caste community Maratha. Most of them live on the hilly terrain of the village. The Thakar in particular reside on the hilltop in Mhaswandi which is a hill village. The dwellings of Bhil from Kangar are slightly on the upper part of the village. Kangar is on the plains. The minimum land that they own is rainfed. Very few have irrigated land. Few tribal families live on the plains along the main road. They possess irrigated land. However, it is less in quantity than their counterparts, the Maratha. The Thakar and the Bhil earn their livelihood from varied means, the agriculture and agricultural labour work, other labour work like that under the Employment Generation Scheme (EGS), seasonal labour work obtained from migration to outside places, hunting, fishing and forest food gathering. The Thakar go for hunting and fishing in the forest. They hunt for monitor lizard (*ghorpad),* rabbit and crabs. They also collect honey (*madh*). Meat, fish, forest food forms a part of the diet of Thakar children.

The cultural characteristics of the tribal community here like migration, search of varied means of livelihood loosens their contact with watershed development inputs as a whole and probably this could be one of the reasons behind the slow pace of change observed among the tribal community as compared to the caste.

The Lamani are a nomadic tribal community in Maharashtra leading a settled life in the village Kalas Pimpri. They had settled in the village between 1975 to 1988. The Lamani have come from a place, Manikdoh. They purchased some land in the village and started practising agriculture. The land that they possess is of rainfed type and is in patches spread over at distance *(jeerayati aaple yede wakde pasarleli jamin hayte)*. They practice agriculture during the rains, are involved in labour work in coal mines in Karnataka and

Rajasthan and sugarcane harvest outside the village. The couples migrate out together leaving behind their children with grandparents. The small children up to age one to one and a half years are carried along to the places of migration. The Lamani women are bold and articulate. After the initial rapport building, the women discussed in the focus group with a confidence.

Utility of Concept Points Scale

'Scale' as a research tool is long been used for measurement of certain observations in psychology. Here the scale has been used to measure health culture change.

The concept points scale measuring change can be developed on the basis of the development inputs given to a community. The information given to a community can be numbered. The information points can be plotted on a scale numbering from one to infinite depending upon the information/inputs given to a community. While assessing change in a community or a particular group, change can be measured in terms of adherence to the points plotted on the scale which would mean adherence to the development inputs given to that group. More the points expressed are consistent with development inputs/scientific literature more the change.

The concept points scale has been of great use in the present study. The points could be plotted on the scale or can be shown in the form of a matrix. It has helped in understanding health culture change that is change in perceptions of a community. As the scale was helpful in our study, we thought that it should be shared. It could be useful in development research, qualitative research and when working on a complex comparative study, may be with comparisons at more than two levels. Social change researchers and development practitioners may use, try and test if they think it is of worth. The scale needs to be tested and reviewed critically and the results shared.

With the concept points scale, here it was easier to understand where a particular community or a particular study group of interest stood so far as social/cultural change and the desired development impacts are considered. It is a simple tool, simple to use, has been simple to present qualitative data analysis, is inexpensive and gives accuracy. The tool however demands professional maturity in qualitative research: data collection, grasping the contents, analysis and interpretation.

Once the measure and direction of change are known it helps improve upon the development work. It is hoped that the scale would also prove useful in informing the status of the development projects to the stakeholders (the community, the donors, the partners and others).

The concept points scale measuring culture change can be expanded further and it is left to the practitioners, either research or development, to use it the way they want and mould it to their work. It is just a small attempt made here to make use of scale. It is just putting a thought, a starting point to carry out work on scale. What should the scale be called may differ from study to study. Social scales need to be developed to measure the impacts of development. Many different types of concept scales can be developed depending on the objective of the study.

Note: The copyright of the Concept Points Scale as developed here to measure and present health culture change rests with the author of the book.

6

Community Practices Related to Child Nutrition and Health

The topic on child health practices covers data on practices related to birth weight, infant and child feeding, immunisation and the underlying concepts and beliefs of the people. So far as the community practices are concerned, it is found that most practices are similar across the three comparative groups, viz., villages, Mhaswandi (WSD+GM), Kangar (WSD) and Kalas Pimpri (No Input). The differences are prominent across ethnic groups, the caste and the tribe. Hence, they are presented accordingly.

The similarities and differences across the comparative groups and those across the communities are discussed together in the following text.

Suggestions for policy making and development inputs are given in bullets in the discussions.

Birth Weight

The issue of birth weight here is closely related with the delivery practices prevalent in the community and therefore it is discussed at length.

Among both the Maratha and the tribal communities, both home and hospital deliveries are prevalent, with a slight difference in preference.

Among the Maratha, the hospital deliveries are more common than the home deliveries. Observation indicates that the Maratha from Mhaswandi village have more hospital deliveries than the other villages. The Maratha women from Mhaswandi have mentioned fear of baby death due to complications in delivery. Observation indicates problems related to children like still birth, premature deliveries, low birth weight are existent and comparatively more among the Maratha. There are not many Maratha women who can perform or assist a woman in her delivery. The elderly Maratha women like a health promoter, the *Anganwadi* helper who have received government health centre's training in conducting delivery do not take the risk of performing or assisting in deliveries.

In case of few Maratha from all villages, where the deliveries are conducted at home, they are assisted by a doctor either from or the nearby village. The doctor is called on phone. He gives an injection to hasten the delivery. In case of complication as judged by him he refers the case to the private doctors in larger villages or taluka places, in a range of 10–20 km. The doctor in the village is also called at home for giving a tetanus toxide (TT) injection to the woman. The doctor takes the birth weight of the child and refers the low birth weight cases to taluka level hospitals.

In general, the Maratha women from all villages prefer private hospitals for delivery. The Maratha women from Mhaswandi particularly spoke of preference for big private hospitals as they are open for 24 hours. Such hospitals are available at district or taluka places or semi-urban locality in a range of 10–50 km with facility for cesarean section. Government health centres, although closeby are not preferred for delivery as the women said that they are small and closed at night.

The *Anganwadi* workers motivate the women to get delivered in hospital for safe delivery. The material for delivery is safe in hospital.

If an undernourished child is born, it gets appropriate services. It is kept under medical supervision and its growth is monitored.

In case of home deliveries the birth weight of the child does not get recorded. It is recorded when the *Anganwadi* workers come to know about the birth of the child. According to the *Anganwadi* worker, it takes about fifteen days to come to her knowledge about a delivery in the village. The delay can be due to following reasons:

The *Anganwadi* worker mentioned that as they do not get the information of home delivery immediately, the birth weight cannot be recorded. It also happens that the day when a home delivery takes place the *Anganwadi* worker is not available in the village at the same time. Even if a home delivery is immediately known to the *Anganwadi* worker she can attend to the baby and note down the weight of the child residing in a place that is close to *Anganwadi*. There is difficulty in immediate rushing to far off places like to the hamlets situated far off from *Anganwadi* and located on hilly terrain. The weight of the home delivered child is taken on any day when the *Anganwadi* worker comes to know about the home delivery. She comes to know during her home visits or from a fellow villager during informal discussions. The weight of the child in most cases is taken for the first time when the child is brought to the *Anganwadi* centre for vaccination or during the *Anganwadi* teacher's home visits. In case of home delivery, the people take the woman and the child to the hospital for TT injection. During that time the weight of the baby is taken and noted as birth weight in *Anganwadi* records.

All communities understand birth weight as an indicator of the child's health status. This understanding partly comes from their day-to-day experience. Most men and women mentioned that an underweight baby is kept in incubator *(kachet thevtat)*. Babies with birth weight less than 2.5 kg are considered as weak. More weight means the baby is good in health (*kami bharla manje asaktapana yeto. Jast bharle manje bal bara aahe*). However, no conscious efforts are taken by the community to get the birth weight known and recorded in case of home deliveries. They perceive that if the birth weight is more than 3 kg, there is no problem and the woman receives early discharge from the hospital.

Only those women whose deliveries had taken place in the hospital were able to tell the birth weight of their child.

Earlier the *Anganwadi* workers took weight of the child every three months as mentioned by them in the village Mhaswandi. Since the work on growth monitoring had started, they work in coordination with the health promoters of SAMPADA TRUST and record the weight of the child every month.

Among the Thakar from Mhaswandi and the Bhil from Kangar most of the deliveries take place naturally, carried out either by the woman herself (*naysargik padhatine*) or assisted by an old woman, (*Aaji*), a Traditional Birth Attendant (TBA) from tribal community. The tribal people from all villages believe that they do not need a doctor (*amchya mahila gharich hotat, kahi tras hot naahi, aamchya mahilancha he aasach chalta*).

The tribal women expressed fear in going to hospital for delivery (*ghabartat bai manse*). Because the doctors get wild if the women shout in labour pain (*jasta tras hovayla lagla bai mansala delivery cha tar mug te doctor khavaltat*).

The Lamani women from Kalas Pimpri mentioned about performance of delivery on their own *(sauta balatin jhale)*. In some cases, the delivery has been assisted by a professionally trained person, the auxiliary nurse and midwife (ANM) of government Primary Health Centre, conducting deliveries in her quarter. The people call a private vehicle and reach the ANM's place for delivery. The ANM charges for the delivery. She records the birth weight of the child.

The birth weight is not usually taken when delivered at home. If the delivery is assisted by a health professional, the birth weight is recorded. Among the few tribal women whose delivery was conducted in a private hospital, they were not able to tell about their children's birth weight. When asked, the tribal women mentioned that it ranges from 1 to 2.5 kg and there are few children with birth weight between 1 to 1.5 kg.

It is a perception among the tribal community, specifically the Thakar and the Bhil that more the weight of the baby, more difficult the delivery. The belief is that the woman who can eat more in pregnancy *(jya baila anna jasta chaltay, mang tee bharapur jevati. Mag potamadhye mul dabun rahatay.)*, eats more, her child gets pressed in the stomach. The woman who does not eat much in pregnancy her baby "grows big", meaning thereby that it also grows well (*aani jya baila nahin na jevan chalet teecha mul motha hotai*). So the woman should take less and light food (*asa thodkyat khayacha*) like rice and dal, fruits, chapatti. She should not eat bread *(Bhakar nahi khayachi)*. If the child weighs more, the delivery is difficult. It was observed that in actual practice too a seven months pregnant Thakar woman did not eat anything in the afternoon. She did not feel hungry. Her diet consisted of half *bajra* bread in the morning and available seasonal fruits, like custard apple *(sitaphal)*, mango *(amba)*, *ber (bor)*, *jamun (jambhul)*, *karvand*, etc.

Discussion

Although all communities believe that the health of a child can be judged on the basis of birth weight, in practice they are not keen in knowing the birth weight of the child. No conscious efforts are made either by the *Anganwadi*, the health promoters, or by the community to get the birth weight known and record it even where home delivery takes place. Hence, it appears that people are not so conscious about the birth weight of the child which is a key indicator of the nutritional and the health status of the child.

Observation indicates that adolescent marriages are prevalent among the tribal communities from Mhaswandi and Kangar except the Lamani from Kalas Pimpri. Girls of age thirteen to fourteen are married and start having children by age fifteen. While among the Lamani girls continuing with higher secondary level education and interested in studying further were seen.

The Thakar women themselves conduct their own delivery or the delivery is done by a traditional birth attendant from their community. As culturally they are used to conducting their own delivery, they take care to avoid complications during delivery, that is, by not increasing the weight of the baby. The underlying notion is that with increase in weight of the baby, the baby will not get delivered easily and naturally and the baby may "stuck up" leading to complications and causing risk to the mother. This is probably the thought behind their food consumption practices followed during pregnancy. The health of the mother is given more importance than the health of the child among the Thakar. Such is not the case with regard to Lamani. There are few Lamani women who conduct deliveries on their own. Some take the assistance of a professional (a medically trained person, an ANM). The pregnant women eat according to their appetite and not restrict themselves like the Thakar. This difference in the two different types of tribal communities is probably related to cultural variation. Traditionally, the Lamani is a nomadic tribal community, although settled down at Kalas Pimpri. Probably the nomadic way of life involves in take of more energy and therefore less restrictions on food consumption during pregnancy and secondly the considerations of food availability from region to region do not permit much food restrictions as those found among the settled tribes.

It is found that there is a cultural change in preferences for treatment seeking. Preference for large, private hospitals with facilities round the clock by women from a prosperous village Mhaswandi belonging to WSD+GM study group was found.

- It is essential to rise awareness regarding the importance of recording of birth weight in all communities as it is not given much importance by the community and/or by the NGO.
- It is essential to change the outlook of the tribal community towards birth weight of the child.
- The Traditional Birth Attendants (TBA) of the tribal community need to be equipped to perform deliveries of women giving birth to good normal weight and healthy babies. This should be a part of the curriculum of government training of the TBAs, from the point of view of reducing malnutrition among the tribal community.

Infant Feeding Practices

Infant feeding practices include practices followed from birth until age six months.

Irrespective of ethnic groups and the comparative study groups, it is a common practice among all the communities to feed the new born baby with items like boiled sugar water or plain water or milk of cow/goat for the first three to four days after birth. The water or milk is fed with the help of a piece of cotton (*varcha dudh pajaycha. Garam karun gar karaycha ani asa bolyane pajaycha*). The tribal communities too follow the same practices. Colostrum is discarded (*aapli madhye mus rahate na tevadha kadhayacha. Kharab asataya na tyachyamule ti kadhun phekun dyava lagataya. Nista ek dhar kadhun phekaycha)*. It is squeezed out on the baby's cloth *(balacha dhudka asato aapla sopan sadha, ranti, tya dhudkyavar kadhito)* or on a cow dung cake *(shenachi gauri)*. This is the traditional practice and they could not tell the reason for discarding it on the cow dung cake. The belief is so strong that a woman reported her baby been given glucose water after her caesarean delivery in the hospital. It is believed that lactation is not initiated until then. Once started, breast feeding is continued until two years.

Some Maratha women mentioned about feeding honey to the new born immediately after birth. The belief is it helps in language fluency of the child.

Colostrum (*pahile dudh)* is discarded among all communities. On the fourth day during breast feeding, the initial flow of milk is squeezed out as it is believed to be harmful to the child (*pilun takaycha nantar pajaycha. Aai mhante kharab asta*). The elderly women and the mothers of the delivered women discourage immediate breast feeding. There are different reasons for discarding of colostrum. Colostrum is not considered as normal milk. The implicit understanding about milk is that it is thin and watery. It is believed that the colostrum is not digestible for the babies *(balala khapat nahi na!)*. Pain in breast was pointed out as another reason for the practice. A strong belief is that lactation is initiated after three to four days *(mang pyayalach nahi na rahat)*.

In case of hospital deliveries, the doctors promote immediate breast feeding and the women follow the hospital rule. Some Maratha women reported following the traditional practice of feeding sugar water to the babies under the influence of their family elders and hid the fact from the doctors.

The Maratha women from Mhaswandi (WSD+GM group) and Kangar (WSD) perceive that colostrum *(cheek)* has nutritive elements *(jeevansatva aaste)*. It helps to raise the immunity (*rog pratikarak shamata*) of the child and that it protects the baby from any disease and especially jaundice (*kavil*). They got the information from the doctors. However, in actual behaviour they follow their traditional practices.

Once started, breast feeding continues at least upto two years. The women have reported going to work in the fields three to six months after delivery, depending upon the busy agricultural season. If the farms are close to the house then they return to feed the baby. If the farms are far off, the grandmother who takes care of the child at home feeds external milk (cow/goat milk) to the baby with the help of container and spoon until the arrival of mother. Some women take their children to the field. The child is kept in a cradle (*jholi*) tied to the tree. A grandparent sits beside the child to take care of it. The women feed the child as and when the baby cries. Exclusive breast feeding continues until six to seven months. After completion of six months, they start giving it cow/goat milk. The external milk is given three to four times a day.

The milk is diluted with water. Approximately 25 per cent water is added. Undiluted milk is never given to children. Water is always added, even a single drop is fine, said the people, but a must. There is a strong belief in all communities underlying dilution of milk. That is, boiling of pure milk leads to burning of tits of the milch animals (*gaychi, bakrichi kaas jalate*) affecting milk production. That is why water is added to milk. Secondly, direct consumption of pure milk causes indigestion problems is the belief.

Discussion

The Maratha women have explained the importance of colostrum *(cheek)* in detail, correlating it with immunity that is health of the child (Refer to previous section on perception change). As against this the Maratha of Kalas Pimpri village did not say anything about the importance of colostrum. They did not appear to be so keen on feeding colostrum to the child after birth.

Anthropological findings suggest colostrum feeding as a natural practice among the tribal people. Discarding of colostrum is not found in areas with predominant tribal population. A change in this practice is indicative of change under mainstreaming of the tribal culture. A personal view point is that a deeper understanding of tribal culture indicates that the tribal community believes and usually follows the nature's cycle. They have a harmonious relationship with nature. If the calf clings to the mother cow after birth, a tribal would hardly prevent it from the action. In caste farming communities, milk production is commercialised where the colostrum of cows and buffaloes is made available for human consumption. For example, sweet dish made out of it is sold in the market. The fodder is made available to the calf as a substitute for the cow's milk. Culture contact with the mainstream communities has probably brought about a change in the practices of the tribal communities discontinuing the practice of breast feeding immediately after birth.

Child Feeding Practices

Child feeding practices include those practices that are followed since the child starts having supplementary food.

Among all communities, the children are fed external milk after the completion of six to seven months. Until then they are exclusively breast fed. As the child grows up, after a year or so, some amount of black tea is added in the milk and given *(chaha dudh)*. The idea is that it tastes sweet and children like to drink it. The milk is diluted with water. The proportion of water to milk is about 25 per cent among the Maratha and the tribal community, the Lamani.

The proportion of water to milk is more among the tribal Thakar and the Bhil, which is around 35-40%. As compared to the Thakar and the Bhil, the Lamani enjoy a better economic status partly due to cash crop production of cotton in large amounts and partly from the huge profits from employment in coal mines in Karnataka and Rajasthan where they migrate.

Weaning takes place in the ninth month or so. The child is introduced to a soft *(mau)* food item to start with. A glucose biscuit is the weaning food popular among all communities. Tiger brand glucose biscuits are popular all over the villages. Items like *dal* (watery part), cooked vegetable water, finely grained and cooked rice meshed with *dal*, or with milk, more finely cooked rice *(bhatachi pej)* or a piece of fruit are also given to the child of age around nine months or so as a supplementary food. The people stress on weaning with soft food items and begin when the child naturally demands food, when the child expresses his desire to eat. Items like biscuit, a piece of fruit are given in the hands of the children to find out if they wish to eat, if they have developed the appetite *(biscuit kimva phalacha tukda dyaycha hatamadhye)*. So the introduction of solid food is initiated only after a trial test.

The underlying concept is that the child is not able to chew and swallow hard food items properly and the food particles may get stuck up causing risk to the child. This fear leads to weaning at a later age of that around ninth month or later. The usual pattern is the child who has completed a year gradually starts eating solid food items. The bread *(Bhakri)* made up of *bajra* or *jowar* is crushed and meshed with cooked vegetable *(kalwan), dal* or milk and some sugar is added. The meshed food is locally called as *kala/malida*

The Maratha from Kalas Pimpri village said that children do not eat food for a year or so *(ek ek varsha khat nahit mula. Pot nahi bharlyavar charayala lagayacha)*. When the child is hungry and is not satisfied after breast feed the women understand that it is the time to start with supplementary solid food. This time is around a year or so. They understand the development of appetite from the child's behaviour that is from crying. They said that if a child is hungry it cries even after breast feed. And the age of about a year is such that the appetite has increased. A woman gave an example of her child who did not eat until the completion of two years *(Aamcha he porga kiti don varshacha hoistavar khat navhta)*. The usual pattern is that at the age around a year, as the child expresses his desire to eat, the child is fed bread made up of *bajra* or *jowar* or chapatti made up of wheat meshed in cooked vegetable (*kalwan*) or *dal* or in milk with some sugar added.

It was observed that the children among all communities are often given biscuits when they cry. A child of age one and a half years eats six to eight biscuits in a day. The children were found given *bhel* and other junk food items brought from the market in tribal families. Feeding of junk food items to children was observed in Mhaswandi. "A biscuit stops the child from crying and it is convenient for the mothers occupied with farm work" was the common voice of all women.

- Awareness raising inputs regarding weaning practices, frequency of supplementary food to be given to children, health impacts of junk food items for children are needed so as to improve the current nutritional practices and status of children.

Discussion

Due to the fear of risk, weaning begins in the ninth month or so in Mhaswandi. Weaning is much too late in Kalas Pimpri (beyond one year). Due to occupation with farm work the children are left to eat on their own. They are given a food item in their hands. There is no monitoring of his/her eating behaviour: may that be about the frequency, the quantity of food eaten, thrown or wasted. It is left to the child and to nature. The underlying concept could be that child eats depending upon his own appetite and digestion capacity. Therefore, the parent does not intervene and allows the child to eat as much as and whenever wanted. Economic prosperity from watershed development seems to permit expenditure on junk food which needs attention of development sector for health interventions. These practices could have serious implications for the nutritional status of the children.

- Growth inputs need to concentrate efforts on changing weaning practices through appropriate means.

Immunisation

It was observed that people belonging to all the communities across all comparative groups are keen about their children's immunisation. They believe that immunisation prevents the children from handicaps and disease. Most women from all communities gave the information on the kinds of doses given to their children during interviews and focus group discussions. They take their children for immunisation to the *Anganwadi* centre on the scheduled day as communicated by the *Anganwadi* workers during their home visit. All important announcements are made in the village. The village temples have a mike or a drum used for making announcements. The immunisation day is announced in the village.

For the Maratha, child health is considered as the responsibility of the women. The Maratha women expressed that it is the duty of the mothers to take their children for immunisation because only the women know about the birth date of their children and that they can look after their crying children better than men. In few cases when the mother is too busy with the farm work, the father or the grandparent and in case of migration outside, when the child is left with the grandparent at home, the grandparent takes the child for immunisation but usually never misses the scheduled dose. As against this, it was observed that for the tribals, child health was not a subject of division of roles and responsibilities. Both the parents share the responsibility of taking their children for immunisation.

It is important to note that on the whole there is a good awareness and keenness among all to get the child immunised against diseases and is one of the factors contributing to child health irrespective of watershed development and growth inputs. The ICDS (*Anganwadi)* and the government Primary Health Centres (PHC) have significantly contributed to the successful coverage of children for immunisation.

The *Anganwadi* workers maintain the record of the children immunised. Those who do not attend, they are given doses by the ANM of PHC by paying home visit. This is the usual pattern.

The problem is only with the migrated children. In the cases of children who migrate out along with their parents, it is generally believed that they take their scheduled doses at the place of migration. However, no record is kept regarding the immunisation a child receives in the period of migration with parents.

- It would be worthwhile to maintain a record of immunisation of children who migrate out with their parents for a longer duration. The mothers should be made aware of the immunisation of their children that is due in the period of migration.

7

Nutritional Status of Children

Qualitative data analysis reveals a change in the perceptions of the community and a difference in change between the three study groups. On the whole, health consciousness of the community with watershed development and growth inputs is on the rise. However, the data on community practices shows a variation from the data on community perceptions. It indicates not much change in behaviour and not much difference in the comparative study groups. This indicates that the change in health consciousness is not reflected in the change in health behaviour.

As the health consciousness is greater in WSD+GM study group which sets the group apart from the others in terms of cultural change, a research query was to understand if the raised health consciousness is reflected in the nutritional status of children. It was to understand whether the results of qualitative study show any consistency with nutritional status of children. The *Anganwadi* periodically assesses the nutritional status of children on the basis of weight for age. The weights are recorded using a spring balance weighing machine. The weights are plotted on the growth charts developed by ICDS for assessment.

Anganwadi keeps a record of nutritional status of children 0–6 years in all villages. The data on nutritional status of children 0–6 years were collected from the *Anganwadi* centres of three villages belonging to three comparative study groups, Mhaswandi (WSD+GM), Kangar (WSD) and Kalas Pimpri (No Input). Statistical significance of the data is tested with Epi Info ™ Version 3.5.1. software (http://www.cdc.gov/epiinfo/downloads.htm.) just to have a rough idea.

In Mhaswandi and Kalas Pimpri, there is one *Anganwadi* centre while in Kangar there are four *Anganwadi* centres. The data from all four *Anganwadi* centres were collected in Kangar. The data are presented as follows:

Table 13

Nutritional Status of Children 0–6 Years Indicating Normal Status and Gradewise Malnutrition Across Three Comparative Groups/Villages

Village (Comparative Group)	*Total Number of Children 0–6 Years*	*Category*	*Normal*	*Malnutrition Grades*			
				I	*II*	*III*	*IV*
Mhaswandi (WSD+GM)	146	Number	90	53	2	1	0
		Percentage	61.64	36.30	1.38	0.68	0
Kangar (WSD)	505	Number	258	197	46	3	1
		Percentage	51.09	39.01	9.11	0.59	0.20
Kalas Pimpri (No Input)	118	Number	62	43	10	3	0
		Percentage	52.54	36.45	8.47	2.54	0

The above results are put in graphical form as shown in Figure 8.

Interpretation of Table 13 and Figure 8: More number of children with normal nutritional status are found in watershed village where

Figure 8

Nutritional Status Compared Gradewise Across Three Comparative Groups

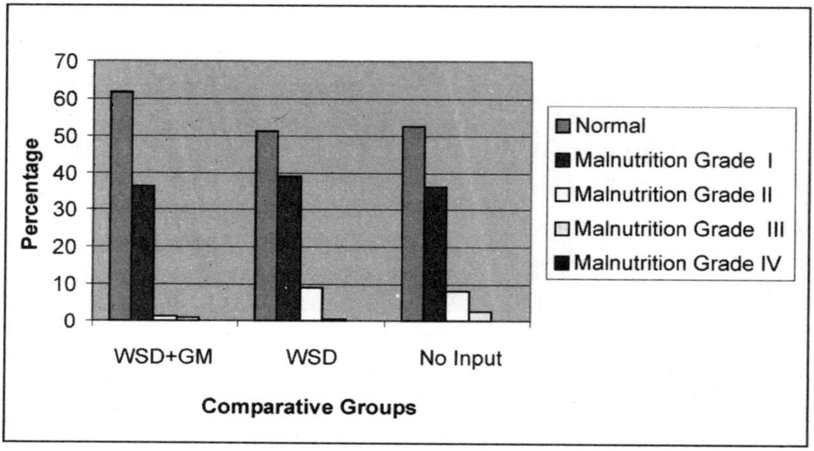

participatory growth monitoring inputs are initiated. The number of normal children in the village with watershed development input alone is slightly lower or could be said as almost equal to the village with No Inputs. The results points out that growth monitoring inputs seem to be contributing in enhancing the nutritional status of children. The difference between WSD+GM and WSD group is statistically significant.

Watershed development may lead to agriculture development and economic prosperity but that does not mean it leads to enhanced nutritional status is what the result indicates.

Other observations indicate more number of children belonging to malnutrition grade II in Kangar, a village with watershed development inputs and more to malnutrition grade III from Kalas Pimpri, a village with No Inputs of watershed development and growth monitoring.

Table 14

Nutritional Status of Children 0–6 Years Indicating Normal Status and Moderate and Severe Forms of Malnutrition Across Three Comparative Groups/Villages

Village (Comparative Group)	*Total Number of Children 0–6 Years*	*Category*	*Normal*	*Malnutrition*	
				Moderate (I+II)	*Severe (III+IV)*
Mhaswandi (WSD+GM)	146	Number	90	55	1
		Percentage	61.64	37.68	0.68
Kangar (WSD)	505	Number	258	243	4
		Percentage	51.09	48.12	0.79
Kalas Pimpri (No Input)	118	Number	62	53	3
		Percentage	52.54	44.92	2.54

The above results are put in diagrammatic form as shown in Figure 9.

Figure 9

Nutritional Status Compared as per its Severity Across Three Comparative Groups

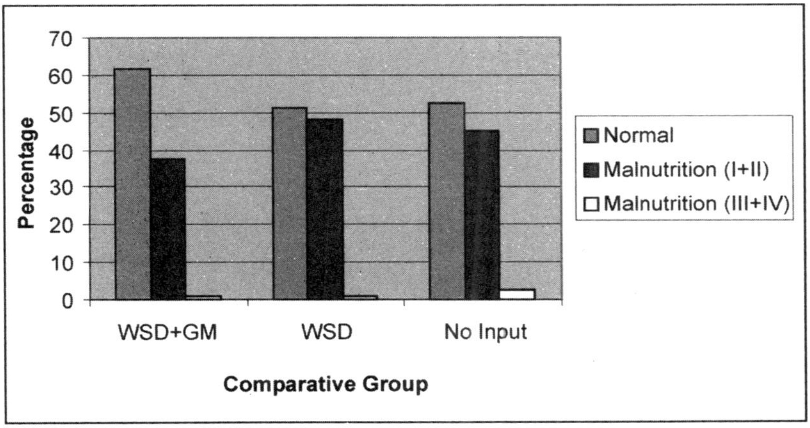

Table 15

Sexwise Nutritional Status of Children 0–6 Years Indicating Normal Status and Gradewise Malnutrition Across Three Comparative Groups/Villages

Village and Comparative Group	Sex	Total Number 0–6 Years	Nutritional Status									
			Grades									
			I		II		III		IV		Normal	
			N	%	N	%	N	%	N	%	N	%
Mhaswandi (WSD+ GM)	Girls	66	43	65.15	21	31.81	1	1.52	1	1.52	0	0
	Boys	80	47	58.75	32	40.00	1	1.25	0	0	0	0
	Total	146										
Kangar (WSD)	Girls	253	111	43.87	110	43.48	28	11.06	3	1.19	1	0.40
	Boys	252	147	58.33	87	34.51	18	7.14	0	0	0	0
	Total	505										
Kalas Pimpri (No Input)	Girls	57	26	45.62	21	36.84	7	12.28	3	5.26	0	0
	Boys	61	36	59.01	22	36.07	3	4.92	0	0	0	0
	Total	118										

Note: Symbol 'N' denotes number and % as percentage.

The ICDS classifies grades I and II as moderate malnutrition and grades III and IV as severe forms. Hence, the terms 'moderate' and 'severe' used here are based on ICDS classification (Ref: http://wcd.nic.in/).

Interpretation of Table 14 and Figure 10: There is moderate malnutrition in children in Kangar (WSD group) and severe malnutrition in children in Kalas Pimpri (No Inputs). Moderate malnutrition in Mhaswandi (WSD+GM) is comparatively less. Here too the difference between WSD+GM and WSD group is statistically significant again indicating the significance of participatory growth monitoring.

The above results are put in graphical form below as shown in Figure 10.

Interpretation of Table 15 and Figure 10: When the nutritional status is observed sexwise, a similar picture is obtained as was observed for the total children irrespective of sex.

Figure 10

Nutritional Status Compared Gradewise Across Gender

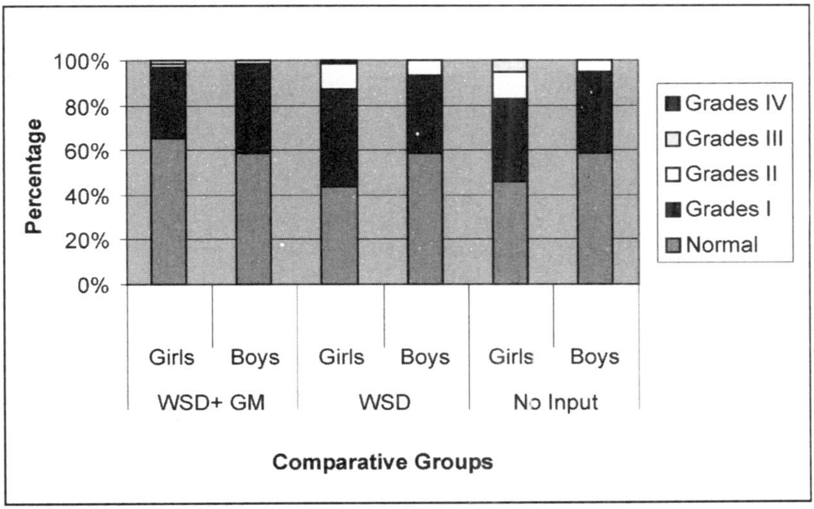

More number of girls from village Mhaswandi (WSD+GM) group have normal nutritional status than other villages and study groups. The difference in the nutritional status of girls in WSD+GM group with nutritional status of girls from WSD and No Input groups is statistically significant.

The WSD+GM group has more number of girls than boys with a better nutritional status. As against this, the situation is reverse in the other two study groups. More number of boys than girls have normal nutritional status in WSD and No Input groups. The difference between the nutritional status of boys and girls in WSD group is statistically significant.

Table 16

Sexwise Nutritional Status of Children 0–6 Years Indicating Normal Nutritional Status and Moderate and Severe Forms of Malnourishment Across Three Comparative Groups/Villages

Village and Comparative Group	*Sex*	*Total Number 0–6 Years*	*Nutritional Status*					
			Normal		*Malnutrition*			
			N	*%*	*Moderate*		*Severe*	
					N	*%*	*N*	*%*
Mhaswandi (WSD+GM)	Girls	66	43	65.15	22	33.33	1	1.52
	Boys	80	47	58.75	33	41.25	0	0
	Total	146						
Kangar (WSD)	Girls	253	111	43.87	138	54.55	4	1.58
	Boys	252	147	58.33	105	41.67	0	0
	Total	505						
Kalas Pimpri (No Input)	Girls	57	26	45.62	28	49.12	3	5.26
	Boys	61	36	59.01	25	40.99	0	0
	Total	118						

Figure 11

Nutritional Status Compared as per its Severity Across Gender

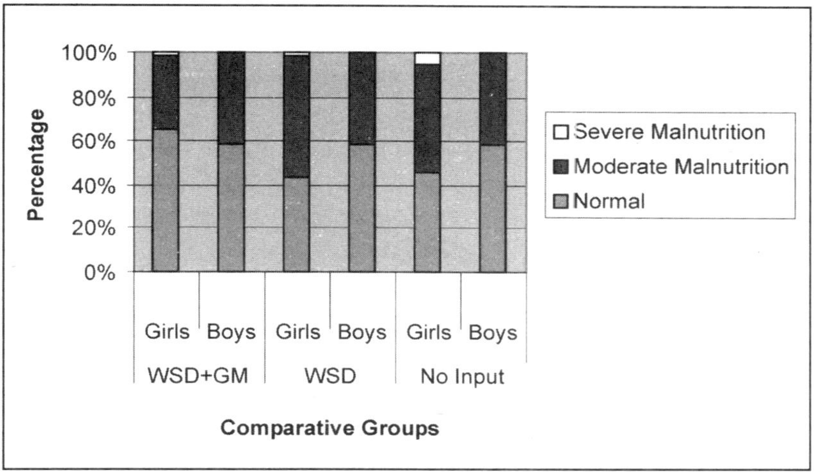

The results are put in graphical form as shown in Figure 11.

Interpretation of Table 16 and Figure 11: When moderate and severe forms of nutritional status are observed sex wise, a similar picture is obtained as was observed for the total children irrespective of sex.

More number of girls than boys have normal nutritional status in Mhaswandi (WSD+GM group).

More number of girls than boys are moderately malnourished in Kangar (WSD group). The difference is statistically significant.

More number of girls than boys are severely malnourished in Kalas Pimpri (No Input group).

Discussion

The results of nutritional status of children indicate the following:

1. When participatory growth monitoring is initiated in watershed villages, it helps in improving the nutritional status of all

children, girls and boys. The nutritional status of girl child is better than boys.

2. In watershed villages, where participatory growth monitoring is not initiated and there is only a single input of watershed development, malnutrition exists but in moderate form.
3. When there are no inputs of watershed development and growth monitoring, malnutrition exists in severe form.

8

Gender Issues

Watershed development inputs (watershed development and participatory growth monitoring) are closely related to the lives of rural women in various ways: to help develop women institutions, to help facilitate economic growth impacting on women, to help them have their own savings, to help them reflect and improve upon their children's nutritional status. Therefore, the issues surrounding women are important from the point of view of understanding the impacts of watershed development inputs and the social change. It was not the objective of the study to gather data related to the topic on gender issues. However, during the stay in the villages, observations and informal conversations with people brought to light gender differences as perceived and practised by the communities. Gender differences impact upon the nutrition and health status of children and the girl child in particular. Gender equity is of great concern in development sector and programmes catering to women are specifically planned for their overall development. A thought was to have some space in the book reserved to discuss issues surrounding the lives of women.

The chapter discusses the observations of the researcher. It is found that across all the three comparative groups, there is a gender difference found in the perceptions and practices. The work concentrates on community perceptions towards the girl child and its implications on child nutrition.

It is found that gender perceptions of the caste communities, the Maratha and that of the Lamani tribal community (referred to as group 1 here) are very much identical, while, the view points of tribal communities, viz., the Thakar and the Bhil (referred to as group 2 here), are identical. The qualitative data are presented below:

Among caste communities and the Lamani, there is greater preference for sons if they already have had a girl child. In focus group discussions with women and informal conversations with people from their communities, a voice of son preference was intensely expressed.

People do not explicitly state gender difference but in actual practices are found to treat the girls differently from boys and have laid more stress on having a male child.

Age at Marriage and Education of Girls

Group 1

A Maratha woman mentioned that the girls are married young (*mulicha lavkar lagna hote*). The Maratha girls are married off by the time they are 17–18 years. The search for the boy begins as soon as the girl attains puberty. Age 17–18 for girls is considered as appropriate in the villages and a right age for marriage as compared to the marriage age in the cities. The perception is that if the girl is not married at the appropriate age, there is difficulty in finding a boy. There is societal pressure for marriage of girls as soon as she attains puberty. Among the Maratha, the girls are allowed to continue with their school education until matriculation if they show interest in continuation of studies. Married young women of age 22–25 were found to have completed school education from seventh to tenth standard.

In Mhaswandi (WSD+GM) group, it was observed that a woman from other castes, a *Teli* caste (a community of oil pressers in the traditional Hindu caste system) who is a member of SMS *(Samyukta Mahila Samiti)* and a health promoter, has undergone several training programmes related to health intervention and other development inputs, encouraged her daughter to continue with higher studies, that is at least upto graduation and stand on her feet. Factors like membership of a village development institution, exposure to outside influences and interactions with groups wherein the importance of education, health, development is stressed, own experiences have played a significant role in decision-making for her daughter's education and economic independence.

The Lamani perceive that the girls have to be married by the age of eighteen, as they grow old very fast while the boys can be married by the age of 24–25. Lamani girls were found to continue with education beyond twelfth standard. There are few houses of Lamani in village Kalas Pimpri. A girl who secured first class in tenth and twelfth standard was encouraged to continue with her studies.

Group 2

Among the tribal communities, the Thakar and the Bhil, early marriages are common. The age at marriage for girls is around 14–15 years. By the time the girl is fifteen, she has at least one child born. Illiteracy is widespread. However, few young women of age twenty had completed education upto seventh standard and that is the maximum the tribal women have in the study villages.

Perception towards Girl Child

During one of our other studies on clinical examination of children 0–5 years by a doctor, few Maratha men came forward and inquired with the doctor if she could perform a sonography, a medical examination of one of their relatives who was pregnant. They wanted to determine the sex of the child as the family desired for a boy child. When asked for the reasons for son's preference, they expressed that a girl on growing old gets married off and is not able to look after her parents. The girl is considered as the property of others and not of her parents, while the boy remains with his parents life long *(Mulgi mothi jhali ki lagna houn jail. Mulgi aai wadilankade baghel ka? Mulga pahije. Mulga gharchach)*. They said that there is a government norm that sanctions for one girl and one boy child. *(Shasanane ek mulga, ek mulgi pass kela aahe)*.

The Lamani believe that the girl is the property of someone else and not her parents. The girls are valued for their compassion and generosity as against boys. They said that the girl takes care of parents more than what the son can take for which a daughter is needed in the family. However, the family is believed to be incomplete without a male child.

During the discussions with the Thakar and the Bhil, a question related to the difference in girl and boy child was taken with a surprise. The women said that there is no difference between a girl and a boy. There is no difference in their health and sickness. That since birth they are just as what they are, the girl and a boy *(jalamlya pasun haye tasech haye!)*. A Bhil informant specifically pointed out that their community is different from others. That the birth of a girl child is welcomed among them unlike the other communities. That theirs is a unique community who are kind at heart. That they can never differentiate between a girl and a boy child (*amcha samaj eakmev aasa dayalu aahe. Amcha samaj mula balancha babtit ase karu shaknar nahi*). He said that even if a couple has had four daughters they would never think of aborting the next girl child.

Perceptions regarding Gender Differences in Child Health

The Maratha and the Lamani were of the view point that the girls are healthier than boys. The girls eat anything that is given to them and do not cry or crave for a particular thing. The boys are perceived as weak in comparison to girl children. The Lamani mentioned about pampering of the boy child. "The boys need more of pampering otherwise they grow weak"*(mulala jeev lavaya lagato na jasta!)*. They said, "The boys cry a lot *(mule khup radati)*. They must be pampered". Referring to a family where there are four daughters, the fifth time a boy child was born, the boy child was pampered. The woman asked the researcher whom would they pamper a girl or a boy child? *(mulila jasta jeev lavten kaa mulala?)*. It was noted from the discussions and observations that pampering takes place in terms of feeding the children. They opined that the girls can be given anything to eat and they keep quiet *(mulila kasa aahe, kahi bi dila tari khanar)*. The girls look bigger than the boys from childhood to puberty. The boys are not so. That their food demands have to be fulfiled.

The qualitative data shows consistency with results of nutritional status based on *Anganwadi* records as discussed in the previous chapter. The nutritional status of girls is poorer than that of boys in villages, Kangar and Kalas Pimpri.

In a Maratha community, in the *Anganwadi* teacher's house we observed that a girl of age around 12–13 was not given food until all the elderly members in the house finished eating. It is a norm that women and girls in the house should feed first and then have food for themselves.

There is a regular practice of abortion of the girl child said the *Anganwadi* worker from village Mhaswandi. This is particularly so with the third girl child onwards. Sex determination tests help in decision-making of abortion. Medical examination like a sonography is used to determine the sex of the foetus. If a family has two daughters already, the third pregnancy onwards sex determination is done and the decision of abortion of the female foetus is taken. The tests are performed in nearby towns and at *taluka* places. This accounts for a lower sex ratio found in Mhaswandi as against other villages.

The *Anganwadi* workers belong to caste community Maratha. In conversations with the *Anganwadi* workers from village Mhaswandi, the importance of the male child was highlighted as follows. The *Anganwadi* helper spoke in high voice, laying stress on the issue.

The people perceive that the government norms do not permit to have third child. They shared their experience that the name of the third child is not recorded on ration card, not taken on *grampanchayat* records, not enrolled in *Anganwadi*. When there are two daughters, there is urge for having at least one son. If the third one is a girl child, it is aborted. "What is the use of the third girl child born when the Government rules don't permit?" questioned the *Anganwadi* helper (*sarkarkadun tisarya mulacha adhikar nahi na! Ata jar mala don mulich aahet. Mag aata tisari mulagi jhali, mulaga nahiye, mag aata tisarya mulila adhikar nahiye tar tisarya mulicha upayog kay. Mulaga jhalyawar to gharchach mhatlyavar thevitat. Mulga nasan tar mulgi kothun yete*?). "There are families which have five to six daughters. Then what would they do? Wouldn't they undergo abortion in such case?" kept asking the *Anganwadi* worker (*pach puri jhalyat tari tyancha vishay toch. Aata tya garbhapat kartil nahitar kay kartil*).

The *Anganwadi* worker questioned the researcher in high tone asking that those who have four and five daughters with no son, wouldn't they abort the girl child. Her questioning in high voice gives some indications of her implicit agreement to the practices of sex selective abortion.

The Thakar and the Bhil do not perceive a difference in the girl and the boy child in any way *(mulat aan mulit kahi pharakach vatat nahi)*.

Sex Ratio

The gender perceptions and practices among the caste community Maratha communities have severe implications on sex ratio and consequently on maternal and child nutrition and health. The urge for son causes the women to conceive early and frequently as the family is perceived to be complete only after having a son, thereby minimising the period of birth interval. Gender difference and discrimination is not much an issue among the tribal Thakar and Bhil communities in the study villages. The prevalent practice of bride price among the tribal accords a value and superior status to tribal girls and women.

The following is a table giving the total and sex wise population and calculated sex ratio of the children 0–5 years for the study villages.

Table 17

Children 0–5 Years

Study Group	*Village*	*Total Children*	*Total Numbers*		*Sex Ratio*
			Boys	*Girls*	
WSD+GM	Mhaswandi	122	62	60	968
WSD	Kangar	421	206	215	1,044
No Input	Kalas Pimpri	147	68	79	1,162

Figure 12

Sex Ratios Compared

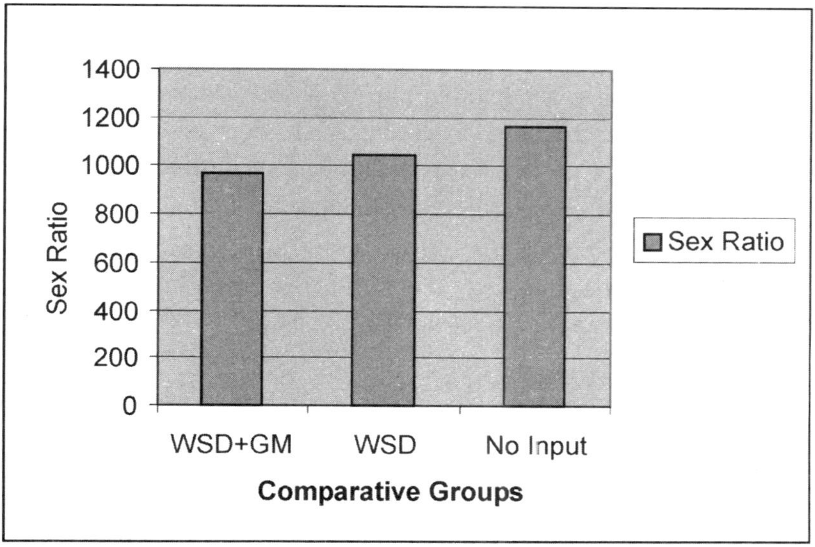

Interpretation of Table 17 and *Figure 12:* The sex ratio for children 0–5 years for the villages seems to be decreasing with the increase in development inputs. The highest sex ratio is in Kalas Pimpri, having zero input while Mhaswandi, with two inputs, watershed development and growth inputs has the lowest.

The economic prosperity from watershed development seems to permit expenditure on sex selection in Mhaswandi.

Child Mortality

A gender difference is observed in child mortality in three villages discussed from data as follows. The data on child mortality of children 0–5 years is taken for one year, January to December 2007. It is based on *Anganwadi* records cross-checked with information from the village elders.

Mhaswandi: In Mhaswandi, there have been three cases of child deaths among the Maratha. Two of which had been still births of girl child. In one case, the cause of still birth was delayed delivery (*dilevery la jast vel lagla mhanun mul potat varla*). In another case, the foetal sounds were poor (*pandhara divas pasun balache thoke kami hote*). Both the deliveries took place in the hospital. The third case was due to brain injury to the child (*mendu la maar lagla*). The child was a boy, age about three years old. There is a case of infant mortality of a girl child among the Thakar in the past one year. The cause of death was prolonged fever that was not treated. The people said that the treatment was not affordable for the family.

Kangar: In Kangar, there are two cases of premature deaths among the Maratha children in one year of age between 0–1 month, a boy and a girl. There is no case of mortality among the Bhil in the past one year.

Kalas Pimpri: In Kalas Pimpri, there has been a case of death of girl child among the Maratha at the age of three. There is no case of child mortality among the Lamani. The Lamani mentioned that during illness they immediately give treatment to the child.

The data on child mortality is presented in the table as follows:

Table 18

Child Mortality

Village and Comparative Group	*Total Number (Age 0–5 Years)*			*Number of Reported Death Cases Community and Sexwise*				
	Children	*Girls*	*Boys*	*Caste*		*Tribe*		*Total*
				Girls	*Boys*	*Girls*	*Boys*	
Mhaswandi (WSD+GM)	122	60	62	2	1	1	0	4
Kangar (WSD)	421	215	206	1	1	0	0	2
Kalas Pimpri (No Input)	147	79	68	1	0	0	0	1

Figure 13

Child Mortality

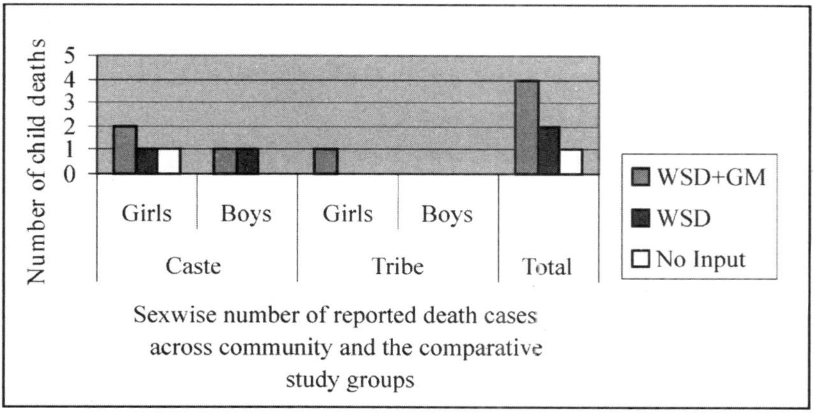

The table indicates more number of child deaths in Mhaswandi. The social change which has come about in the village through watershed development and participatory growth monitoring inputs shows a reverse trend when visualised from the point of view of child mortality.

More number of death cases are found among the Maratha and more are of girl child. The total number of deaths of girl child is five out of seven. Still birth and premature deliveries are more common among the Maratha. Why should there be still births and premature deliveries of the girl child remains a question. They are prominent in a caste community whose perceptions denote a low value for daughters. The data points out relationship with the perceptions of the community towards girl child. Although the data are based on small sample size, it gives some pointers of the trend existing and poses questions. The tribal communities have less number of girl child deaths. The culture of tribal communities accords a higher status to women. The prevalence of the practice of bride price and gender discrimination not perceived are indicative of the value the women have in the tribal culture. Among the Lamani the situation is slightly different. Practices of bride price and dowry both are prevalent. The practice of bride price is on the verge of diminishing. But they are encouraging girls to seek higher education and women completed schooling have been found. Therefore, among the Lamani it can be said that the women are still accorded a higher position.

The Maratha women themselves do not value women. It was noted that women's health is not of much concern for the prosperous Maratha. Pregnancy care is lacking. The women do not seek treatment for illnesses occurring during pregnancy like severe vomiting continued for nine months, indigestion problems were left untreated. Pregnancy care in terms of prevention and treatment of illness, consumption of nutrition supplements like iron and folic acid tablets is neglected by the Maratha women as observed in Kangar. Although they implicitly understand the relationship of mother's health with that of the child as observed from community perceptions, the understanding is not translated in actual practice.

If one looks at the chapters on nutritional status of children and gender issues in holism, a fact that comes to light is that the sex ratio for children 0–5 years is lower in WSD+GM group but on the nutrition front, there are more number of girls having better nutritional status than boys in this group. On the other hand, in WSD group, the sex ratio for children 0–5 years is much better but on the nutrition front, there are more number of girls with poor nutritional status than boys and this difference is statistically significant. These phenomena are taking place in watershed villages. It could be said that gender discrimination is present in both the groups. In one it takes place before birth and in the other it takes place after birth that is in childhood period of age 0–5 years. Where it takes place before birth, it is occurring in an area with growth inputs, which could be the reason that there is no gender discrimination in nutritional status of children observed.

These findings have policy implications indicating that participatory growth monitoring inputs should make headway with providing development inputs not only after birth but much before the birth of a child. Watershed development too needs to progress much further getting sensitive towards gender concerns and walking the path leading towards gender equity.

9

Linkages

When the research data and interpretations are viewed in totality across the three comparative groups, it gives us some important indications discussed as follows. All the interpretations made are personal view points based on minimal experience discussed at length. They are open for discussion, critical comments and new thoughts.

Community Issues in Development

Tribal Community

My own anthropological academic and field experience with the tribal communities from different parts of Maharashtra: Thane, Dhulia, Nasik districts and from parts of Orissa: Koraput district and literature review has formed a picture of a tribal community and their culture in mind. The tribal people have lived and some are still living in isolation. Isolation is relative and differs from one tribal society to another. The tribal societies are one in contact with outside world and in the process of change. However, if we define isolation in terms of connectivity of the tribal with the outside world, their integration with the mainstream culture, local to global, we will find that many are still living in isolation. The term 'isolation' used in connection with the tribal world needs a redefinition in the world where global culture has come into existence. Therefore, isolation still remains a feature of the tribal culture as compared to caste societies. The tribal people are primarily found as inhabiting the forest and hilly areas. They have small, scattered land holdings. The cultivation is hardly enough to sustain them. Their economy is characterised as subsistence economy. Technology is of small scale and there is less awareness and exposure to new methods of cultivation and practices (Pandit, 2006). As there is no surplus generated, they are alienated from the market economy. For livelihood they depend on many different sources: forest food gathering, fishing and hunting, labour work, besides cultivation if and wherever possible. Due to the diminishing nature of the forests, they are in a horrible state of living. They have lost their original means of livelihood. Their alienation from the market forces has been a cause of their exploitation in the hands of the traders. The deprivation that results out of the many causes contributes to rising malnourishment of their children, the group that is crucial in terms of human health and development.

Having this understanding at the background, when the tribal communities are studied from the point of view of watershed development and other inputs, one finds a social change among the tribals. The data on tribal community presented and discussed indicates that a change from subsistence to surplus economy is taking place among the tribal small scale farmers. The examples from data about tribal small farmers cultivating cash crops like tomato and selling it in market, the recognised expertise of the tribal people in tomato cultivation, growing crops like wheat, garlic, onion to meet livelihood needs, growth of women institutions like tribal women in self-help groups point out a social and cultural change in flux and heading forward in the change process.

There is a change from subsistence to surplus economy, links with market forces are getting established, rise in health consciousness as against the belief in supernatural in tribal culture, awareness about child nutrition and health helping in better management of the nutritional problems of the children.

We do not have the exact figures of nutritional status for the tribal children as a separate category as the objective was not to do a community wise analysis. Therefore, we cannot comment on the nutritional status of the tribal children here. However, the overall figures from watershed development and participatory growth monitoring inputs (WSD+GM) group show a greater degree of normal nutritional status than the other groups.

Caste Community

The Hindu caste community Maratha is an agricultural caste community. The social structure and culture tuned to agriculture and allied growth and market understanding. Patriarchal culture, big joint families, strong preference for son, the male members, who hold land on their names, who undertake the responsibility of seed sowing and not the women is a culture different from the tribal. The caste Hindu are concentrated in the plains which provide them with fertile cultivable land. Unlike the tribal people, agriculture is mechanised. Maximisation for profit is the motive.

The two structures, the tribe and the caste are at two different cultural ends. As per data, the Maratha have concentrated on onion production. Onion is a commercial crop. The caste economy is heading towards commercialisation. The caste Hindu farmers also show a change in terms of getting commercial benefits from agriculture. There is a change from surplus to commercial economy. Onion production for commercial benefits and the rise in dairies are often associated with watershed development as exemplified in the data. In the study area, the caste community Maratha are dominant numerically, economically and politically. They exercise control over natural resources and have greater role in decision-making for village development. The influence of the dominant caste and the cross-cultural differences in the two communities affect the livelihood and overall development of the tribal greatly when there is ample scope and opportunity for the tribal farmers to progress with watershed development. These would have implications for nutritional and health status of tribal children. Watershed development organisations can play a significant role in intervention by generating the spirit of social equality. The NGO staff particularly should be sensitive towards community issues. Development workers need to look at communities with cultural relativism and value neutrality.

Women and Development?

More Voice

When the data and the discussions of the three comparative groups are seen together, an alarming change is observed. In the group where watershed development and growth monitoring inputs are present, a change that can be measured in terms of greater degree of exposure, knowledge possessed/health consciousness, mobility, articulation, confidence, voicing of opinions, struggle for achieving rights and participation in the village development processes by women is notable. The development in women is found in a village which is comparatively isolated, whose infrastructure is comparatively weak: examples of poor connectivity and services from data are worth-noting. Examples from data like ban on liquor by women, closing down three liquor shops in Mhaswandi, struggle and gain of rights in family property, strength of women institutions and entrepreneurship are sufficient enough to explain the change and the development processes that have set among the women. Although we find this change in women in Mhaswandi which is a watershed village with growth monitoring inputs as against Kangar, the changes that have taken place in women have no connection with growth monitoring inputs. The strength of women institutions is greater in Mhaswandi and the development activities like growth monitoring are a continuation of post-watershed development NGO contacts and inputs which seem to be helpful in strengthening the women institutions.

The strength of women institutions in watershed development and growth inputs group is remarkable in the development processes of the villages. All these forces result in greater awareness of child nutrition and health, raising the nutritional status of children 0–5 years.

The women who exhibit a greater degree of confidence, exposure, awareness even in the absence of development inputs, for example as was observed in a small population group of Lamani women is probably due to cultural variation. Lamani is a nomadic tribal community, settled in the village. The women and men migrate for work outside the state to coal mines and earn huge profits. They speak of going for *dhanda* (business) rather than *majuri* (labour) as referred to by other communities. The Lamani women are the exception. Similar change is not found among the Maratha women from No Input group as discussed in the data analysis.

Gender Equity: How Much of Concern?

While the women from Mhaswandi (WSD+GM input group) exhibit a change in behaviour, the confidence and change is not found in voicing the concern for the lower sex ratio in the village. Among all, Mhaswandi shows a declining trend in sex ratio. The child mortality indicates more number of girl child deaths. With economic prosperity, there is more cash at disposal than before permitting expenditure on sex selection. Gender discrimination is a product of patriarchal culture in part. Visaria (2008) points out women themselves have internalised the patriarchal values to such an extent that more sons than daughters are desired. The data illustrate that among the Maratha, the Hindu agriculturist caste, discrimination for girls exists before birth (e.g., Mhaswandi) and continues at the level of feeding as exemplified from the nutritional status differences in Kangar.

Gender issues indicate urgent policy attention. Economic growth as initiated with watershed development and health consciousness as with growth inputs is not accompanied with social development. Gender equity has remained a gray area for watershed development.

A lower sex ratio, gender differences in nutritional status favouring boys, more of girl child mortality would seriously affect the health and overall development of women and children. The raised level of awareness and strong women institutions with watershed development inputs yet need to achieve a lot in terms of gender equity. More voice, more confidence, more knowledge of child nutrition and health is not much contributing to the development of women themselves as gender equity still remains to be achieved.

Child Nutrition

When the community perceptions and the practices are compared together, one finds a great change, a great difference in the perceptions of different communities and change across the comparative groups. The practices however show a different trend. It appears that practices of infant and child feeding are governed by cultural notions of colostrum feeding, breast feeding, weaning and food consumption. In most cases, these practices are followed on the advice of the elderly people. Although there is no agreement about the right age of weaning in the medical and health sciences, weaning as late as one year is likely to lower the nutritional status and growth of the children.

One of the important changes that have occurred in practice is the change in the number of meals. A change from two to three meals in a day is important from the point of view of nutrition and is noticed by the people themselves.

A large change in the perceptions of the women with children 0–5 years in the area of watershed development inputs shows a change in the perceptions of the young population. The change has crept in but yet has to enter into practice, that is, reaching actual behaviour.

Development: The Other Side of the Coin

Watershed development has been criticised here for not actually impacting upon social development. The impacts may be visible on economic and environmental fronts. Watershed development is a technical and not a social intervention and there are limitations to what extent social returns from watershed development can be expected. The technical interventions have social implications and the book discusses in detail social and cultural issues that need to be considered in planning for watershed development programme. Watershed development and other inputs are just an example in case to observe social and cultural changes. However, the results, the analysis, the discussions and considerations apply to the development sector as a whole.

Participatory growth monitoring inputs need to progress much further to increase the nutritional status of children further. The work on which is an ongoing activity and is in the initial stages of implementation. Therefore, it would be too early to expect the increased impact of these inputs at the level of practices and child nutrition status. People's perceptions and practices are hard core of culture and development inputs need to be given some more time to make change visible.

The perceptions have changed as evident from the comparison, which means that culture has changed in part and is in flux. A change can be sustainable only when it becomes a part of culture and the societies are heading towards sustainable change influencing the child nutrition in terms of its enhancement.

So far as the figures of nutritional status of children are concerned, they are based on *Anganwadi* records of children 0–6 years. This data were collected so as to give an idea of the nutritional status of children. as against the results of qualitative study.

Watershed development efforts where growth monitoring inputs are initiated are likely to yield better results as exemplified by child nutrition status.

10

Learnings: Policy and Research Considerations

The chapter highlights learnings for future research and policy considerations for the development sector. The chapter identifies the areas where development inputs are essential. These considerations are applicable to development sector as a whole and watershed development in particular.

Malnutrition (undernourishment) is a grave problem of the society. The study teaches us a lot of lessons. It indicates that nutritional status improves with watershed development inputs (watershed development and growth monitoring associated). These inputs are bringing about a change in the health perceptions of the people and a culture promoting new ideas for child nutrition and health has emerged. If growth inputs are causing an impact, it is worth, rather essential to take efforts on these lines. Women, their local institutions, the self help groups are playing a vital role in child nutrition and the changing culture around it. Because growth monitoring inputs are found giving results in a place where women institutions are strong, where collaboration between village women, the health promoters and the ICDS staff exists. Women institutions are a part of comprehensive and integrated watershed development programmes. Watershed development leads to economic growth. Watershed development therefore is no doubt an important input in the semi-arid and drought-prone rural areas. However, it needs to be strengthened with social and health inputs. It is essential to focus on making the programme more and more effective. Therefore, watershed development, growth monitoring inputs, the strength of women, collaborations between government and non-government workers, community participation should contribute in forming a good strength to fight against the problem of malnutrition.

The study suggests any development input designed for income generation, for socio-economic growth must be associated with social and health inputs for society as a whole and children of age 0–5 years as a priority.

More research is recommended to validate the results and the conclusions made in this study. Qualitative studies offer a rare insight into the process of development. For a change that promotes new ideas, promotes options for better health must not only be measurable but should seep into the minds of the people only after which it can be sustainable.

The development inputs related to growth monitoring of children 0–5 years need to concentrate efforts to raise community awareness in the following areas:

- Significance of birth weight in the health status of the child.
- Significance and promotion of colostrum feeding.
- Weaning: Appropriate age of weaning for rural areas, importance of weaning food in the status of child health.
- Appropriate frequency of supplementary food. Intervention programmes to improve child feeding behaviour need to be designed and implemented.
- Maintenance of record of immunisation of the children migrating out with parents.

The strength of watershed development plus growth inputs with women power is proved. In a place where this strength is prevalent, the status of women is declining measured in terms of sex ratio and gender differences in perceptions as discussed in the study. Malnutrition is decreasing but the sex ratio is slowly declining. When the women institutions are so strong that they have fought to gain human rights (example of struggle of a woman achieving right in family property with the help of women in the village) have failed to fight against the mindset and the structure of the society devaluing the girls, the women power. Rather, they have not realised the gravity of this malady. Watershed development has a great challenge before it to bring about a change in this perspective. It is a complex task for it is a march towards a structural change. The study indicates economic growth (raised income levels) accompanied with watershed development has probably permitted the people to make use of technology of sex selection. An input that will raise the consciousness of the society towards the value of women could be associated with watershed development efforts. Learnings from the thoughts put forward by Visaria would be of great help in this regard. Visaria (2008) suggests that "society's fears and concerns can be addressed through real life stories of girls and women who have made their parents proud of them and the thought that investment in daughters who after marriage are someone else's property is unproductive".

Watershed development needs to be supported with inputs for consciousness raising for valuing girls. Women institutions could be used as the medium to facilitate the change. The consciousness raising should begin with involving women in the task. The status of women so enhanced with these efforts will further enhance the nutritional status of children 0–5 years.

More research is recommended to understand the changing status of woman in the context of watershed development inputs.

The tribal culture is undergoing a change in the context of watershed development inputs interacting with market economy, with credit and savings through self-help groups, forces which were once alien to them. Their background of complete unfamiliarity with market economy may put them at disadvantage. The development organisations can play a significant role in providing inputs related to market transactions for the tribal community. The development staff should be sensitive to the tribal culture and act to mitigate the disadvantages causing hurdles in the processes of tribal development, the development of the poor.

The sensitivity to ethnic composition and community structure is important to be borne in mind for the development sector to understand the needy, the poor and the disadvantaged.

The understanding of issues and cultures of the tribe and the caste groups is important for the overall development of the specific region. For example, the tribal people are at a disadvantage under the influence of dominant caste in the study villages.

There is a strong need of the development inputs raising the negotiation skills of the tribal people on the one hand helping them to voice their concerns and participate in village development processes in "greater manner" and of inputs to the caste society on the other in enhancing their understanding regarding the importance of social equity without which development is incomplete.

The change factors discussed in the book are based on a study of three villages, with one village in one comparative group. The factors which are found to be significant may not be the only factors contributing to change. For example, women's institutions are strong in Mhaswandi and have been a medium for change and the relative success of watershed development. A study with involvement of large number of villages may put forward different other factors. It is pointed out that the strength of the village which contributes to a positive change must be identified and should be further strengthened and used as a medium to facilitate future desired changes in the interest of development.

References

Aube, Judi and Mansour, Mohamed. 1989. Qualitative Community Health Research: A Tunisian Example. *Health Policy and Planning*. 4(3): pp. 244–250.

Bamji, Mahtab S. 2003. Early Nutrition and Health-Indian Perspective. *Current Science*. 85(8): pp. 1137–1142.

Braun, Joachim von et al., Essays 5 to 19. Available online at http://72.14.235.104/search?q=cache:oSMrClJEZvcJ:www.ifpri.org/PUBS/books/ar2003/ar03e.pdf+Agriculture+development+and+chil+nutritionandhl=enandct=clnkandcd=9andgl=in. Accessed on 22 September, 2007.

Brennan, Lance et al., 2004. Infant Feeding Practices and Chronic Child Malnutrition in the Indian States of Karnataka and Uttar Pradesh. *Economics and Human Biology*. 2: pp. 139–158.

Das, Sibabrata. 2008. Childhood Undernutrition: A Comparative Analysis of Scheduled Tribes and others in the mid-Indian Tribal Region. *Social Change*. 38(1): pp. 64-83.

Deshprabhu, Rashmi. 2001. *Child Development and Nutrition Management*. Book Enclave. Jaipur.

D'souza, Marcella. 2001. Participatory Watershed Development: Impacting Undernutrition in Rural India. Available online at http://www.hsph.harvard.edu/takemi/RP189.pdf. Accessed 20 September, 2007.

Epi Info ™ Version 3.5.1. Release date: August 18, 2008. http://www.cdc.gov/epiinfo/downloads.htm. Accessed on 15 December, 2008.

54 millionaires in a village. *Down to Earth*. 16(17). January 16–31, 2008.

Food, Health and Care. 1996, November. UNICEF. New York. Updated edition.

Government of Maharashtra. Letter dated 25 May, 2006. Declared official list of communities. Copy obtained from Social Welfare Department, Pune.

Government of Maharashtra. 2003. Gazetteer of the Bombay Presidency, Ahmednagar District. XVII. Mumbai.

International Institute for Population Sciences (IIPS) and Macro International. 2007. National Family Health Survey, (NFHS)–3, 2005–96, *India: Key Findings*. Mumbai: IIPS.

Jelliffe, D.B. 1966. The Assessment of the Nutritional Status of the Community. World Health Organisation. Geneva.

Kapur, Deeksha et al., 2005, April. Dietary Intake and Growth Pattern of Children 9–36 Months of Age in an Urban Slum in Delhi. *Indian Pediatrics:* pp. 351–356. Available online at http://www.indianpediatrics.net/apr2005/apr–351–356.htm. Accessed on 30 September, 2008.

Karve, Irawati. 1991(reprint). *Hindu Society: An Interpretation*. Deccan College. Pune.

Kerr, John et al., 2002. Watershed Development Projects in India: An Evaluation. International Food Policy Research Report, Washington, D.C. Research Report 127.

Khadilkar, V.V. et al., 2007. IAP Growth Monitoring Guidelines for Children from Birth to 18 years. *Indian Pediatrics*. 44 (March). pp. 187–197.

Kirsten, Johann et al., 1998. Determining the Contribution of Agricultural Production to Household Nutritional Status in KwaZulu-Natal, South Africa. *Development Southern Africa.* 15 (4): pp. 573–587.

Kulkarni, Shaunak and Sarma, Eva Rani. 2001. Nutrition among Karbi Tribe of Assam. *J. Indian Anthrop. Soc.* 36: pp. 223–229.

Kulkarni, Shaunak. 2008, March. Assessment of Nutritional Status of Some Tribal Groups of Maharashtra. CSS Occasional Paper Series. Centre for Social Sciences. University of Pune.

Pandit, Apoorvaa. 2006. *Tribal Culture and Technology*. Inter-India Publications. New Delhi.

Pelto, Gretel H. and Freake, Hedley C. 2003. Social Research in an Integrated Science of Nutrition: Future Directions. *Journal of Nutrition.* American Society for Nutritional Sciences. Bethesda. 133: pp. 1231–1234.

Ramalingaswamy, V. et al., 1996. The Asian Enigma, The Progress of Nations. UNICEF: pp. 11–17.

Reeler, Dough. 2007. Theories of Social Change. CDRA, available online at http://www.cdra.org. Accessed 19 August, 2008.

Roche, Chris. 1999. Impact Assessment for Development Agencies: Learning to Value Change. Oxfam Development Guidelines. Oxfam.

Sampada Trust. October 2005-January 2006. Child Care and Growth Monitoring Project Report, Phase I and II. Ahmednagar.

Samuel, Abraham et al., 2007. Watershed Development in Maharashtra—Present Scenario and Issues for Restructuring the Programme. Technical Report. SOPPECOM. Available online at http://www.indiawaterportal.org/tt/watershed/res/Maharashtra_Watershed_Forward.pdf. Accessed on 30 September, 2008.

Shah, Amita et al., 2004. Benchmark Survey for Impact Assessment of Participatory Watershed Development Projects in India. Gujarat Institute of Development Research. Ahmedabad.

Srinivas, M.N. 1987. *The Dominant Caste and Other Essays.* Oxford University Press. Delhi.

Tylor, E.B. 1924. *Primitive Culture.* 2 Vols. 7th edition. Brentanos. New York.

Available online at www2.truman.edu/~rgraber/cultev/tylor.html.

Venkateshwarlu, C. and Reddy, K.K. 2000. Growth and Nutritional Status of Pre-Schoolers in Different Caste Groups of Rural Villages in Chittoor District of Andhra Pradesh with Special Reference to Income Levels. *J. Indian Anthrop. Soc.* pp. 79–84.

Visaria, Leela. 2008, March 22. Improving the Child Sex Ratio: Role of Policy and Advocacy. *Economic and Political Weekly:* 43 (12 and 13): pp. 34–37.

Watershed Organisation Trust. Mhaswandi Watershed Project. *Watershed Voices Newsletter.*

Watershed Organisation Trust. 1997. Operations Manual for Capacity Building Phase of the Indo-German Watershed Development Programme, Maharashtra, India.

Watershed Organisation Trust. 2005. Drop by Drop: The Story of WOTR.

Website http://ahmednagar.gov.in accessed on 05 November, 2008.

Website http://wcd.nic.in/ Statewise details of classification of nutritional status of children under ICDS scheme as on 28 February, 2008 accessed on 27 November, 2008.

Website http://www.indianexpress.com/news/1–3rd-of-worlds-undernourished-kids-in-india/3754/

Why Prez is visiting Gawdewadi. *The Times of India.* Pune, September 27, 2008. pp. 1–7.

Yadav, Rajiv and Roy, J. 2005, January and July. Nutritional Status of Pre-School Children among Bharia Tribe of Patalkot. *M.P. Tribal Health Bulletin.* 11(1 and 2): pp. 11–16.

Index